From Imaginary Oxymora to Real Polarities and Return

A New Science of Reality

by Dr. Hans-Joachim Rudolph, MD

Microvita Research e.V.

authorHOUSE®

AuthorHouse™
1663 Liberty Drive
Bloomington, IN 47403
www.authorhouse.com
Phone: 1-800-839-8640

Published by AuthorHouse 2/27/2012

ISBN: 978-1-4685-0843-7 (e)
ISBN: 978-1-4685-0844-4 (sc)

Library of Congress Control Number: 2011962125

Table of Contents

Question: What is the silver lining between the quinquelemental universe and idea that is between the initial stage of matter and Microvita?

Answer: Microvita are the initial stage of matter. Although they are matter they are very, very subtle. All of a sudden Microvita are transmuted into matter and matter is transmuted into Microvita. For investigating the initial stage of matter, research is not possible in physical laboratories, but it is possible in the human mind and human soul. Microvitum is closer to the realm of ideas than matter.

Then, what is the silver line of demarcation between matter and idea? Of that silver lining, the outside is matter and the other side, the inner side, is idea. That is, this silver lining is made of the initial stage of matter and the cruder stage of idea. If you consider that the atom is the constituent of matter, likewise idea is the constituent of Microvita.

Prabhat Rainjan Sarkar
19 May 1989, Calcutta

01. On the Priority of Imaginary Space

In medieval times, the nominalists prevailed in a debate that engaged thinkers from Boethius (ca. 480–525) to Ockham (ca. 1288–1348) and beyond. They asserted that only particulars exist and denied that universals are real, i.e. that they exist as entities or beings. In contrast, the idealists maintained that universals are real and exist independently of that on which they might refer (universalia ante res). A third position was presented by the realists, who took the stance that universals are real to the effect that the particulars instantiate the universals (universalia in rebus).

Although the nominalists prevailed in this debate, they were not able to do so in the field of mathematics, where it's generally assumed, down to the present day, that the existence of mathematical entities is basically independent of human minds. Thus humans don't invent, but rather discover mathematics.

Beyond that, renowned contemporary thinkers hold the opinion that not only mathematics, but also human musical, artistic, and aesthetic creativity and appreciation come from contact with the Platonic world of reality (1).

On top of this, the ultimate question has always been whether there are 'universalia ante res' impinging on our destiny and fortune - and if so, where such universals could reside?

The following derivation is designed to give an answer to these questions:

Let M_R be a Minkowski space, which is a 4-dimensional flat Lorentzian manifold,

and let M_C be the complexified M_R of dimensionality 8,

with 4 real $-t_{Re}$ and x_{Re}, y_{Re}, z_{Re}
and 4 imaginary dimensions $-t_{Im}$ and x_{Im}, y_{Im}, z_{Im}.

Then, the standard basis for M_R will be a set of 4 mutually orthogonal vectors $(-e_0, e_1, e_2, e_3)$, such that

$$(-e_0)^2 = (e_1)^2 = (e_2)^2 = (e_3)^2 = +1,$$

and for M_C there will be an additional set of 4 mutually orthogonal vectors $(-i_0, i_1, i_2, i_3)$, such that

$$(-i_0)^2 = (i_1)^2 = (i_2)^2 = (i_3)^2 = -1.$$

Accordingly, each point p_c in M_C can be written as

$$p_c = (-t_{Re}, -t_{Im}, x_{Re}, x_{Im}, y_{Re}, y_{Im}, z_{Re}, z_{Im}) =$$

$$(-e_0\,t_r, -i_0\,t_i, e_1\,x_r, i_1\,x_i, e_2\,y_r, i_2\,y_i, e_3\,z_r, i_3\,z_i) =$$

$$\begin{bmatrix} -et_r - it_i & ex_r + ix_i \\ ey_r + iy_i & ez_r + iz_i \end{bmatrix} =$$

$$e\begin{bmatrix} -t_r & x_r \\ y_r & z_r \end{bmatrix} + i\begin{bmatrix} -t_i & x_i \\ y_i & z_i \end{bmatrix} = \begin{bmatrix} -t_r & x_r \\ y_r & z_r \end{bmatrix} + i\begin{bmatrix} -t_i & x_i \\ y_i & z_i \end{bmatrix}$$

with $t_r, t_i, x_r, x_i, y_r, y_i, z_r, z_i \in \mathbb{R}$,
$e^2 = +1$ and $i^2 = -1$.

Furthermore, the matrix product of two complex points in M_C will be $p_c \times p'_c = P_C \in M_C \Rightarrow$

$$P_C = \begin{bmatrix} T_r & X_r \\ Y_r & Z_r \end{bmatrix} + i \begin{bmatrix} T_i & X_i \\ Y_i & Z_i \end{bmatrix}$$

with (2)

$$T_r = t'_r t_r + y'_r x_r - t'_i t_i - y'_i x_i,$$
$$X_r = -x'_r t_r + z'_r x_r + x'_i t_i - z'_i x_i,$$
$$Y_r = -t'_r y_r + y'_r z_r + t'_i y_i - y'_i z_i,$$
$$Z_r = x'_r y_r + z'_r z_r - x'_i y_i - z'_i z_i,$$
$$T_i = t'_i t_r + y'_i x_r + t'_r t_i + y'_r x_i,$$
$$X_i = -x'_i t_r + z'_i x_r - x'_r t_i + z'_r x_i,$$
$$Y_i = -t'_i y_r + y'_i z_r - t'_r y_i + y'_r z_i,$$
$$Z_i = x'_i y_r + z'_i z_r + x'_r y_i + z'_r z_i.$$

Now, let

$$Re(-t_r - it_i) = Re(x_r + ix_i) = Re(y_r + iy_i) = Re(z_r + iz_i) =$$

$$Re(-t'_r - it'_i) = Re(x'_r + ix'_i) = Re(y'_r + iy'_i) = Re(z'_r + iz'_i) = 0$$

$$\Rightarrow P_C = \begin{bmatrix} T_r & X_r \\ Y_r & Z_r \end{bmatrix} + i \begin{bmatrix} 0 & 0 \\ 0 & 0 \end{bmatrix} = \begin{bmatrix} T_r & X_r \\ Y_r & Z_r \end{bmatrix}$$

with

$$T_r = -t'_i t_i - y'_i x_i,$$
$$X_r = +x'_i t_i - z'_i x_i,$$
$$Y_r = +t'_i y_i - y'_i z_i,$$
$$Z_r = -x'_i y_i - z'_i z_i.$$

Consequently, let M_I be a subgroup of M_C ($M_I \subset M_C$), defined by this premise:

$$Re(-t_r - it_i) = Re(x_r + ix_i) = Re(y_r + iy_i) = Re(z_r + iz_i) =$$

$$Re(-t'_r - it'_i) = Re(x'_r + ix'_i) = Re(y'_r + iy'_i) = Re(z'_r + iz'_i) = 0$$

$\Rightarrow$

The matrix product P_C of imaginary points p_i, $p'_i \in M_I$

is not an element of M_I $\quad$ ($p_i \times p'_i = P_C \notin M_I \subset M_C$),

but an element of M_R $\quad$ ($p_i \times p'_i = P_C \in M_R \subset M_C$).

Now, let अ$^{(1)}$ be the operator,

which allows a mapping from the imaginary to the real

Minkowski space ($M_I \rightarrow M_R$).

For example:

अ$_1 \in M_I$ with

$$Im(-t'_r - it'_i) = Im(x'_r + ix'_i) = Im(y'_r + iy'_i) = Im(z'_r + iz'_i) = -1$$

$\Rightarrow$

$$T_r = X_r = -t_i + x_i,$$
$$Y_r = Z_r = y_i + z_i,$$

अ$_2$ ∈ M$_I$ with

$Im(- t'_r - it'_i) = Im(x'_r + ix'_i) = -1$,
$Im(y'_r + iy'_i) = Im(z'_r + iz'_i) = 0$

=>

$T_r = X_r = - t_i$,
$Y_r = Z_r = y_i$,

अ$_3$ ∈ M$_I$ with

$Im(- t'_r - it'_i) = Im(x'_r + ix'_i) = 0$,
$Im(y'_r + iy'_i) = Im(z'_r + iz'_i) = -1$

=>

$T_r = X_r = x_i$,
$Y_r = Z_r = z_i$,

अ$_4$ ∈ M$_I$ with

$Im(- t'_r - it'_i) = Im(y'_r + iy'_i) = -1$,
$Im(x'_r + ix'_i) = Im(z'_r + iz'_i) = 0$

=>

$T_r = - t_i + x_i$,
$Y_r = y_i + z_i$,
$X_r = Z_r = 0$,

अ$_5$ ∈ M$_I$ with

$Im(- t'_r - it'_i) = Im(y'_r + iy'_i) = 0$,
$Im(x'_r + ix'_i) = Im(z'_r + iz'_i) = -1$,

=>

$$X_r = -t_i + x_i,$$
$$Z_r = y_i + z_i,$$
$$T_r = Y_r = 0,$$

अ$_6$ ∈ M$_I$ with

$$Im(-t'_r - it'_i) = Im(z'_r + iz'_i) = -1,$$
$$Im(y'_r + iy'_i) = Im(x'_r + ix'_i) = 0,$$

=>

$$T_r = -t_i,$$
$$X_r = x_i,$$
$$Y_r = y_i,$$
$$Z_r = z_i,$$

अ$_7$ ∈ M$_I$ with

$$Im(-t'_r - it'_i) = Im(z'_r + iz'_i) = 0,$$
$$Im(y'_r + iy'_i) = Im(x'_r + ix'_i) = -1,$$

=>

$$T_r = x_i,$$
$$X_r = -t_i,$$
$$Y_r = z_i,$$
$$Z_r = y_i.$$

Which means that the invertible matrix

$$\text{अ}_6 = \begin{bmatrix} 0 & 0 \\ 0 & 0 \end{bmatrix} + i \begin{bmatrix} -1 & 0 \\ 0 & -1 \end{bmatrix}$$

$$= \begin{bmatrix} -i & 0 \\ 0 & -i \end{bmatrix}$$

is the operator for mapping the imaginary to the real Minkowski space ($M_I \rightarrow M_R$).

Moreover, we can define an operator allowing time flashes, i.e. a transformation from a past or future imaginary constellation into the real space-time:

Let $t_i = t_r + a$, $\{a \in \mathbb{R} \mid a > -t_r\}$ and $A = t_r / (t_r + a)$ =>

$$\text{अ}_T = \begin{bmatrix} -iA & 0 \\ 0 & -iA \end{bmatrix}$$

is the operator for mapping imaginary past or future constellations into the real present.

[1] अ: Devanagari A for 'Anujiivat'

02. Let there be Light

Now, let E_R be a 4-dimensional Euclidean space,

and let E_C be the complexified E_R of dimensionality 8,

with 4 real $\quad\quad\quad\quad\quad\quad\quad$ w_{Re}, x_{Re}, y_{Re} and z_{Re},
and 4 imaginary dimensions $\quad$ w_{Im}, x_{Im}, y_{Im} and z_{Im}.

Then, the standard basis for E_R will be a set of 4 mutually orthogonal vectors (e_0, e_1, e_2, e_3), such that

$$(e_0)^2 = (e_1)^2 = (e_2)^2 = (e_3)^2 = +1,$$

and for E_C there will be an additional set of 4 mutually orthogonal vectors (i_0, i_1, i_2, i_3), such that

$$(i_0)^2 = (i_1)^2 = (i_2)^2 = (i_3)^2 = -1.$$

Accordingly, each point p_C in E_C can be written as

$$p_C = (w_{Re}, w_{Im}, x_{Re}, x_{Im}, y_{Re}, y_{Im}, z_{Re}, z_{Im}) =$$

$$(e_0\,w_r, i_0\,w_i, e_1\,x_r, i_1\,x_i, e_2\,y_r, i_2\,y_i, e_3\,z_r, i_3\,z_i) =$$

$$\begin{bmatrix} ew_r + iw_i & ex_r + ix_i \\ ey_r + iy_i & ez_r + iz_i \end{bmatrix} =$$

$$e\begin{bmatrix} w_r & x_r \\ y_r & z_r \end{bmatrix} + i\begin{bmatrix} w_i & x_i \\ y_i & z_i \end{bmatrix} = \begin{bmatrix} w_r & x_r \\ y_r & z_r \end{bmatrix} + i\begin{bmatrix} w_i & x_i \\ y_i & z_i \end{bmatrix}$$

with w_r, w_i, x_r, x_i, y_r, y_i, z_r, $z_i \in \mathbb{R}$,
$e^2 = +1$ and $i^2 = -1$.

Furthermore, the matrix product of two complex points in E_c will be $p_c \times p'_c = P_c \in E_c \implies$

$$P_c = \begin{bmatrix} W_r & X_r \\ Y_r & Z_r \end{bmatrix} + i \begin{bmatrix} W_i & X_i \\ Y_i & Z_i \end{bmatrix}$$

with (2)

$$W_r = w'_r w_r + y'_r x_r - w'_i w_i - y'_i x_i ,$$

$$X_r = x'_r w_r + z'_r x_r - x'_i w_i - z'_i x_i ,$$

$$Y_r = w'_r y_r + y'_r z_r - w'_i y_i - y'_i z_i ,$$

$$Z_r = x'_r y_r + z'_r z_r - x'_i y_i - z'_i z_i ,$$

$$W_i = w'_i w_r + y'_i x_r + w'_r w_i + y'_r x_i ,$$

$$X_i = x'_i w_r + z'_i x_r + x'_r w_i + z'_r x_i ,$$

$$Y_i = w'_i y_r + y'_i z_r + w'_r y_i + y'_r z_i ,$$

$$Z_i = x'_i y_r + z'_i z_r + x'_r y_i + z'_r z_i .$$

Now, let

$$Re(w_r + iw_i) = Re(x_r + ix_i) = Re(y_r + iy_i) = Re(z_r + iz_i) =$$

$$Re(w'_r + iw'_i) = Re(x'_r + ix'_i) = Re(y'_r + iy'_i) = Re(z'_r + iz'_i) = 0$$

$$\implies P_c = \begin{bmatrix} W_r & X_r \\ Y_r & Z_r \end{bmatrix} + i \begin{bmatrix} 0 & 0 \\ 0 & 0 \end{bmatrix} = \begin{bmatrix} W_r & X_r \\ Y_r & Z_r \end{bmatrix}$$

with

$$W_r = - w'_i w_i - y'_i x_i ,$$

$$X_r = - x'_i w_i - z'_i x_i ,$$

$$Y_r = - w'_i y_i - y'_i z_i ,$$

$$Z_r = - x'_i y_i - z'_i z_i .$$

Consequently, let E_I be a subgroup of E_C ($E_I \subset E_C$), defined by this premise:

$$Re(w_r + iw_i) = Re(x_r + ix_i) = Re(y_r + iy_i) = Re(z_r + iz_i) =$$

$$Re(w'_r + iw'_i) = Re(x'_r + ix'_i) = Re(y'_r + iy'_i) = Re(z'_r + iz'_i) = 0$$

=>

The matrix product P_C of imaginary points p_I, $p'_I \in E_I$

is not an element of E_I ($p_I \times p'_I = P_C \notin E_I \subset E_C$),

but an element of E_R ($p_I \times p'_I = P_C \in E_R \subset E_C$).

Now, let अ$_c \in E_I$ be a non-invertible (singular) matrix with

$$Im(w'_r + iw'_i) = Im(y'_r + iy'_i) = -1,$$
$$Im(x'_r + ix'_i) = Im(z'_r + iz'_i) = 0$$

=>

$$W_r = w_i + x_i,$$
$$Y_r = y_i + z_i,$$
$$X_r = Z_r = 0.$$

And let $p_I \in E_I$ be

$$\begin{bmatrix} 0 & 0 \\ 0 & 0 \end{bmatrix} + i \begin{bmatrix} w_i & x_i \\ y_i & z_i \end{bmatrix}$$

with the 'form factors'

$w_i = \lambda$, $x_i = -\lambda$; $y_i = +1$, $z_i = -1$ (or vice versa)

$\Rightarrow$ p_i x अ$_c$ =

$$\begin{bmatrix} +\lambda - \lambda & 0 \\ +1 - 1 & 0 \end{bmatrix} = \begin{bmatrix} +\lambda & 0 \\ +1 & 0 \end{bmatrix} + \begin{bmatrix} -\lambda & 0 \\ -1 & 0 \end{bmatrix}$$

or

$$\begin{bmatrix} +\lambda - \lambda & 0 \\ +1 - 1 & 0 \end{bmatrix} = \begin{bmatrix} 0 & 0 \\ 0 & 0 \end{bmatrix}.$$

Which can be interpreted as two photons with wavelength λ in opposite directions and clockwise (+1) / counterclock-wise (-1) spin being created by the action of the creation operator अ$_c$ for

a short period

$\Delta t \approx \hbar / 2 \Delta E = \lambda / 4 \pi c$

($\Delta E = h * \nu$, $\hbar = h / 2 \pi$ and $\nu = c / \lambda$).

03. Gluon Pair Production

As mentioned before, the matrix product of two complex points in E_C with

$$\text{Re}(w_r + iw_i) = \text{Re}(x_r + ix_i) = \text{Re}(y_r + iy_i) = \text{Re}(z_r + iz_i) =$$

$$\text{Re}(w'_r + iw'_i) = \text{Re}(x'_r + ix'_i) = \text{Re}(y'_r + iy'_i) = \text{Re}(z'_r + iz'_i) = 0$$

is $\quad P_C = \begin{bmatrix} W_r & X_r \\ Y_r & Z_r \end{bmatrix} + i\begin{bmatrix} 0 & 0 \\ 0 & 0 \end{bmatrix} = \begin{bmatrix} W_r & X_r \\ Y_r & Z_r \end{bmatrix}$

with

$$W_r = -w'_i w_i - y'_i x_i,$$
$$X_r = -x'_i w_i - z'_i x_i,$$
$$Y_r = -w'_i y_i - y'_i z_i,$$
$$Z_r = -x'_i y_i - z'_i z_i.$$

Again, let अ$_C \in E_I$ be the creation operator with

$$\text{Im}(w'_r + iw'_i) = \text{Im}(y'_r + iy'_i) = -1,$$
$$\text{Im}(x'_r + ix'_i) = \text{Im}(z'_r + iz'_i) = 0$$

=>

$$W_r = w_i + x_i,$$
$$Y_r = y_i + z_i,$$
$$X_r = Z_r = 0.$$

Now let $p_i \in E_i$ be

$$\begin{bmatrix} 0 & 0 \\ 0 & 0 \end{bmatrix} + i \begin{bmatrix} w_i & x_i \\ y_i & z_i \end{bmatrix}$$

with the 'charge/form factors',
$w_i = r\bar{b},\ x_i = b\bar{r},$ or
$w_i = r\bar{g},\ x_i = g\bar{r},$ or
$w_i = b\bar{g},\ x_i = g\bar{b},$ or
$w_i = r\bar{r},\ x_i = -b\bar{b}$ for the gluon states[2] and
$y_i = +1,\ z_i = -1$ (or vice versa) for the spin

$\Rightarrow\ p_i \times अ_c =$

$$\begin{bmatrix} r\bar{b} + b\bar{r} & 0 \\ +1 - 1 & 0 \end{bmatrix} = \begin{bmatrix} r\bar{b} & 0 \\ +1 & 0 \end{bmatrix} + \begin{bmatrix} b\bar{r} & 0 \\ -1 & 0 \end{bmatrix}$$

or

$$\begin{bmatrix} r\bar{g} + g\bar{r} & 0 \\ +1 - 1 & 0 \end{bmatrix} = \begin{bmatrix} r\bar{g} & 0 \\ +1 & 0 \end{bmatrix} + \begin{bmatrix} g\bar{r} & 0 \\ -1 & 0 \end{bmatrix}$$

or

$$\begin{bmatrix} b\bar{g} + g\bar{b} & 0 \\ +1 - 1 & 0 \end{bmatrix} = \begin{bmatrix} b\bar{g} & 0 \\ +1 & 0 \end{bmatrix} + \begin{bmatrix} g\bar{b} & 0 \\ -1 & 0 \end{bmatrix}$$

or

$$\begin{bmatrix} r\bar{r} - b\bar{b} & 0 \\ +1 - 1 & 0 \end{bmatrix} = \begin{bmatrix} r\bar{r} & 0 \\ +1 & 0 \end{bmatrix} + \begin{bmatrix} -b\bar{b} & 0 \\ -1 & 0 \end{bmatrix}.$$

However, the Gell-Mann matrices (3) provide another four gluon states; three of them (λ_2, λ_5, λ_7) can be derived as follows:

Let $p_c \in E_c$ be

$$\begin{bmatrix} w_r & x_r \\ 0 & 0 \end{bmatrix} + i \begin{bmatrix} 0 & 0 \\ y_i & z_i \end{bmatrix}$$

with $w_r = r\bar{b}$, $x_r = -b\bar{r}$,
or $w_r = r\bar{g}$, $x_r = -g\bar{r}$,
or $w_r = b\bar{g}$, $x_r = -g\bar{b}$
and $y_i = +1$, $z_i = -1$

$\Rightarrow p_c \times अ_c =$

$$P_c = \begin{bmatrix} 0 & 0 \\ Y_r & Z_r \end{bmatrix} + i \begin{bmatrix} W_i & X_i \\ 0 & 0 \end{bmatrix}$$

with

$W_i = +w'_i w_r + y'_i x_r$,
$Y_r = -w'_i y_i - y'_i z_i$,
$X_i = Z_r = 0$.

$\Rightarrow P_c =$

$$\begin{bmatrix} 0-i(r\bar{b}-b\bar{r}) & 0 \\ +1-1 & 0 \end{bmatrix} = \begin{bmatrix} 0-ir\bar{b} & 0 \\ +1 & 0 \end{bmatrix} + \begin{bmatrix} 0+ib\bar{r} & 0 \\ -1 & 0 \end{bmatrix}$$

or

$$\begin{bmatrix} 0-i(r\bar{g}-g\bar{r}) & 0 \\ +1-1 & 0 \end{bmatrix} = \begin{bmatrix} 0-ir\bar{g} & 0 \\ +1 & 0 \end{bmatrix} + \begin{bmatrix} 0+ig\bar{r} & 0 \\ -1 & 0 \end{bmatrix}$$

or

$$\begin{bmatrix} 0 - i(b\overline{g} - g\overline{b}) & 0 \\ +1 - 1 & 0 \end{bmatrix} = \begin{bmatrix} 0 - ib\overline{g} & 0 \\ +1 & 0 \end{bmatrix} + \begin{bmatrix} 0 + ig\overline{b} & 0 \\ -1 & 0 \end{bmatrix}.$$

The gluon state derived from the Gell-Mann matrix λ_8 needs extra consideration:

Let $p_i \in E_i$ be

$$\begin{bmatrix} 0 & 0 \\ 0 & 0 \end{bmatrix} + i \begin{bmatrix} w_i & x_i \\ y_i & z_i \end{bmatrix}$$

with $w_i = \overline{rr} + b\overline{b}$, $x_i = -g\overline{g} - g\overline{g}$,
or $\quad w_i = \overline{rr} + g\overline{g}$, $x_i = -b\overline{b} - b\overline{b}$,
or $\quad w_i = b\overline{b} + g\overline{g}$, $x_i = -\overline{rr} - \overline{rr}$,
and $y_i = +1$, $z_i = -1$

$=> p_i \times अ_c =$

$$\begin{bmatrix} (\overline{rr} + b\overline{b} - g\overline{g} - g\overline{g}) & 0 \\ +1 - 1 & 0 \end{bmatrix} = \begin{bmatrix} (\overline{rr} + b\overline{b} - g\overline{g}) & 0 \\ +1 & 0 \end{bmatrix} + \begin{bmatrix} -g\overline{g} & 0 \\ -1 & 0 \end{bmatrix}$$

or

$$\begin{bmatrix} (\overline{rr} + b\overline{b} - g\overline{g} - g\overline{g}) & 0 \\ +1 - 1 & 0 \end{bmatrix} = \begin{bmatrix} (\overline{rr} + b\overline{b}) & 0 \\ +1 & 0 \end{bmatrix} + \begin{bmatrix} -g\overline{g} - g\overline{g} & 0 \\ -1 & 0 \end{bmatrix}$$

or

$$\begin{bmatrix} (\overline{rr} + b\overline{b} - g\overline{g} - g\overline{g}) & 0 \\ +1 - 1 & 0 \end{bmatrix} = \begin{bmatrix} (\overline{rr}) & 0 \\ +1 & 0 \end{bmatrix} + \begin{bmatrix} b\overline{b} - g\overline{g} - g\overline{g} & 0 \\ -1 & 0 \end{bmatrix}.$$

Therefore, we get not only two, but six possibilities in this case, which finally yields all the eight gluon states (4):

$$(r\bar{b}+b\bar{r})/\sqrt{2}, \qquad -i(r\bar{b}-b\bar{r})/\sqrt{2},$$
$$(r\bar{g}+g\bar{r})/\sqrt{2}, \qquad -i(r\bar{g}-g\bar{r})/\sqrt{2},$$
$$(b\bar{g}+g\bar{b})/\sqrt{2}, \qquad -i(b\bar{g}-g\bar{b})/\sqrt{2},$$
$$(r\bar{r}-b\bar{b})/\sqrt{2}, \qquad (r\bar{r}+b\bar{b}-2g\bar{g})/\sqrt{6}.$$

[2] $r\bar{b}$ = red anti–blue, $b\bar{r}$ = blue anti–red,
  $r\bar{g}$ = red anti-green, $g\bar{r}$ = green anti-red,
   $b\bar{g}$ = blue anti-green, $g\bar{b}$ = green anti-blue,
    $r\bar{r}$ = red anti-red, $b\bar{b}$ = blue anti-blue, $g\bar{g}$ = green anti-green.

04. Putting Things Together

It has been shown that the described mathematical model provides derivations of space, time and energy from imaginary sources (M_I and E_I). Also, future constellations can be projected into present space-time.

The imaginary precursors of force particles (photons, gluons) combine opposites in spin, propagation direction and color charge into one matrix; they undergo polarization only after being transformed into real space - which means that multiplication with the described creation operator actually amounts to a process of symmetry breaking.

Furthermore, the 8-dimensional spaces M_C and E_C can be integrated into one complex block matrix with

$$Q_C = M_C \oplus E_C = \begin{bmatrix} M_C & 0 \\ 0 & E_C \end{bmatrix}$$

as proposed by Emile Grgin (5). Research is now going on with the objective to find more symmetry breaking creation operators $अ^\backprime_C \in Q_I$.

05. Microvita and the Nucleon Model

In physics, nucleon is the generic term for the proton and neutron. The proton is the subatomic particle with a spin of +½, an electric charge of +1 eV (1.602 × 10⁻¹⁹ coulomb) and a mass of 938.3 MeV/c² (1.673 × 10⁻²⁷ kg); similarly, the neutron has a spin of +½, an electric charge of 0 eV and a mass of 939.6 MeV/c² (1.675 × 10⁻²⁷ kg). Both are composed of three quarks (two up and one down quark in the proton, two down and one up quark in the neutron), held together by the strong nuclear force mediated by gluons (6, 7).

With the Standard Model (8), the nucleons' spin and electric charge are explained by the properties of the up (spin +½, electric charge +⅔) and down quark (spin −½, electric charge −⅓); its mass, however, is accounted by quarks for only 1-2 % (7-20 MeV/c²) (9).

To solve this problem, I assume a cubic lattice with a site-to-site distance (D_S) in the range of 2-3 × 10⁻¹⁸ m:

Each site would exist in two potential energy states: $\frac{E_1 + E_2}{2}$ before and $\frac{-E_1 - E_2}{2}$ after production, or $\frac{-E_1 - E_2}{2}$ before and $\frac{E_1 + E_2}{2}$ after annihilation of a pair of virtual gluons respectively (Figure 1).

According to the time-energy uncertainty relation [$\Delta E \times \Delta t \approx \hbar / 2$ with ΔE = the energy conservation violation and Δt = the duration of this violation] (10), the energy of a pair of virtual gluons is limited to $+E_1 + E_2 \approx \hbar / \Delta t$, which means that virtual gluons with low energies exist longer than those with high energies; additionally, a virtual gluon (having no rest-mass) will have internal momentum during Δt of

$p = E / c$ or $p \leq \hbar / (2 \times \Delta t \times c)$.

Now, with a site-to-site distance of 2-3 millifermi, the number of sites inside the nucleon (N_S) comes to 2.7–9.0 $\times$ 10^{+8} [$(^4/_3 \times \pi \times (1.2 \times 10^{-15})^3) / (D_S)^3$] (11); and, suppose each site produces and annihilates 1.6–5.2 $\times$ 10^{+15} pairs of virtual gluons per second (P_S) (Figure 2), then their total energy adds up to $\leq$ 926 MeV per nucleon [$\hbar \times P_S \times N_S$]; and, since the nucleon as a whole is at rest in its rest frame, its average external momentum becomes zero [with $p = 0$, $E^2 = p^2 \times c^2 + m_0^2 \times c^4$ becomes $E = m_0 \times c^2$ and $m_0 = E / c^2$] (3), so that this energy can be fully accounted to fill the nucleon's mass gap of 920-931 MeV/c^2.

But there is something more: According to the latest reviews in particle physics, the quarks account for maximal 60% of the nucleon's total spin, with the remainder being derived from the gluons' spins (Σ_G) and orbital movements (L_G) (12). Therefore, I assume a rotation of the described lattice, due to which the produced virtual gluons get orbital momenta, resulting in an average circulation around the nucleon's spin axis.

Here the question arises whether and how this relates to the concept of Microvita.

In 'Neo-Ethics of Multilateral Salvation', Prabhat Rainjan Sarkar, the propounder of Microvita theory, writes (1987): *"Let us consider the case of the structure of a carbon atom. ... The difference between two atoms is mainly one of nuclear difference. ... Billions of Microvita produce a single carbon atom. That is why it cannot be said that everything comes from carbon atoms. Rather, the carbon atoms come from Microvita. Not only carbon atoms, but all other kinds of atoms are the creation of Microvita."* (13)

So, a lattice with billions of sites would fit into this scenario. But what should these sites be made of?

On May 19th 1979, Sarkar said (14): "*In the beginning, I should say, the supreme causal factor was one. 'Eko'ham bahus'ya'm' - 'I am one, I shall become many.' In the beginning there was only one entity and there was nothing else. This entity, Brahma, comprises two aspects: the cognitive and the operative aspect. These two aspects together constitute one entity. That was the state where there was no scope of expression. So in that state, ultimately, Parama Purus'a determined, 'I am one today, but let Me be many.' 'At the moment I am one, let Me be many.' When this Sam'kalpa was aroused in Its mind, the cosmic mind began.*" A few days later (May 22nd 1979), he specified (15): "*Brahma is a composite of Purus'a and Prakrti. The name for the cognitive aspect is in the masculine gender, and the name for the operative aspect in the feminine gender. But 'Brahma' is in the neutral gender.*" And then, he explained in 'Ananda Marga Philosophy in a Nutshell' (16): "*When the three principles of Prakrti -- sentient, mutative and static -- are in perfect equilibrium and equipoise, Prakrti is said to be lying dormant in Her primordial state ... But when the three principles lose their equilibrium and equipoise, the attributive expression manifests. In this state, when Prakrti is committed to start the creation, (of course within the limited sphere of the infinite body of Purus'a, the supreme cognitive principle), She is called Ma'ya', (the supreme creative principle). The different living beings, the creepers, trees, plants and animals, and this entire world of unique diversities that we witness around us, are all created by Ma'ya'. However, it is only with the expressed will and permission of Parama Purus'a that the vast creation occurs."*

Now, could "this entire world of unique diversities" be created by one Ma'ya' only, or did She also have to become many? We already understood that Parama Purus'a, the cognitive aspect of the supreme causal factor, determined to be many. But if this 'aspect' decides to be many, the other 'aspect' necessarily has to multiply as well. Therefore, I conclude that the fundamental lattice is made of nothing but the multiplied cognitive and operative aspects of the supreme causal factor 'Brahma' (Figure 3).

Microvita and the multiplied aspects of the supreme causal factor

Next to this, the question arises, whether the fundamental lattice (made of the multiplied cognitive and operative aspects of Brahma) is identical with the lattice described for the nucleon model. According to quantum physics, the notions of space break down at distances shorter than the Planck length $(1,6 \times 10^{-35}$ m) (17). So, at site-to-site distances below Planck length, the sites are indistinguishable; in other words, they are in a state of unity, with the sentient, mutative and static principles in 'perfect equilibrium and equipoise'. At distances above Planck length, however, they remain separate, allowing the manifestation of 'attributive expressions'.

From this perspective it becomes clear that the fundamental lattice cannot be identical with the lattice described for the nucleon model. Rather, at a scale of magnitude $\leq 10^{-18}$ m, a single site of pair production and annihilation will contain at least the fundamental lattice with a site-to-site distance in the range of Planck length. Therefore we can say that Microvita float in the aspects of the supreme causal factor, just like quarks and electrons (at another scale of magnitude) float in a sea of virtual gluons and photons respectively.

Nucleic Microvita - what are they actually all about ?

Although Microvita are certainly not the supreme causal factors, they do emanate (18) from its multiplied cognitive and operative aspects: In the atomic nucleus they represent the sites of virtual gluon pair production and annihilation, thereby constituting 98-99% of its mass - and, being embedded in a rotating lattice, they additionally contribute to the nucleons' spin structure.

(3) Actually, there are two solutions to Einstein's energy-momentum-mass equation:

$$E_1 = + \sqrt{p^2 \times c^2 + m_0^2 \times c^4} \text{ and}$$
$$E_2 = - \sqrt{p^2 \times c^2 + m_0^2 \times c^4}.$$

In the 1930s the second one was considered to be unphysical. However, Luigi Fantappié found an interpretation in 1941, which allowed the negative solution to produce *anticipated waves*, diverging from the future to the past (see Chapter 18).

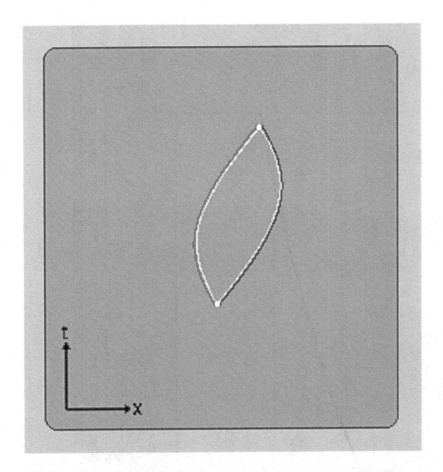

Figure 1: Space-time diagram of a pair of virtual gluons.

Appendix

In terms of Feynman diagrams, the presented calculation is based on four-gluon vertices. If, however, the gluon pair production and annihilation would be based on three-gluon vertices only, P_S would have to be corrected by a factor $k = 4/3$. So, depending on the ratio of three- to four-gluon vertices, its real value will be $k \times P_S$, with $1 \leq k \leq 4/3$.

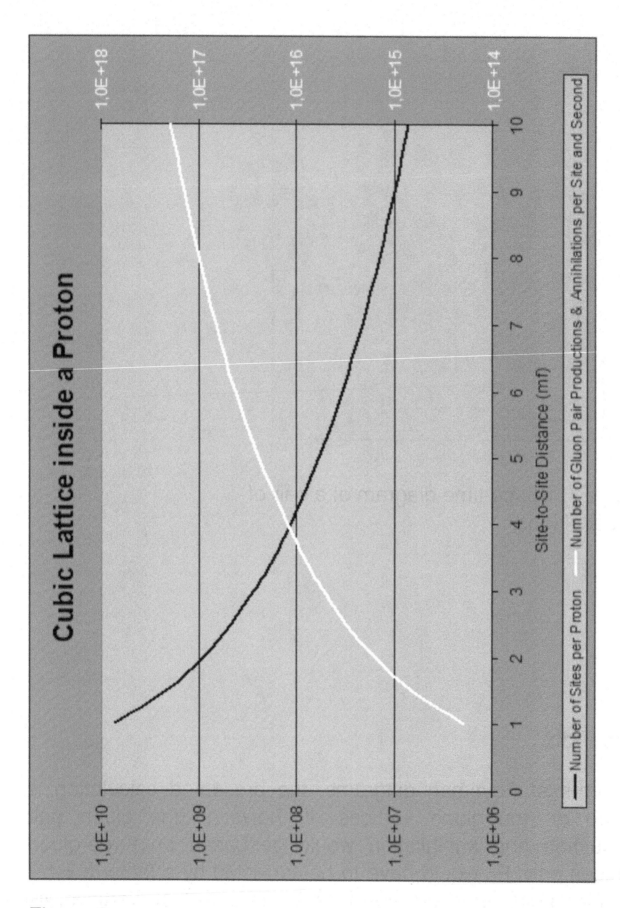

Figure 2

Figure 3: Fundamental lattice made of the 'multiplied cognitive and operative aspects of the supreme causal factor'.

06. The two Faces of Microvita

Figure 1 can be represented as a complex 4 x 4 matrix calculus.

Step 1:

Let $q_l{}^\complement$ and $अ^\complement_c$ be complex block matrices of Q_l with

$$Q_l = \begin{bmatrix} M_l & 0 \\ 0 & E_l \end{bmatrix}$$

$$p_l = \begin{bmatrix} 0-it_i & 0+ix_i \\ 0+iy_i & 0+iz_i \end{bmatrix} \in M_l$$

$$p'_l = \begin{bmatrix} 0+ia_i & 0+ib_i \\ 0+ic_i & 0+id_i \end{bmatrix} \in E_l$$

and the modified creation operator as a singular matrix

$$अ^\complement_c = \begin{bmatrix} 0-2i & 0 & 0 & 0 \\ 0 & 0-2i & 0 & 0 \\ 0 & 0 & 0-i & 0 \\ 0 & 0 & 0-i & 0 \end{bmatrix} \in Q_l$$

$$\Rightarrow q_l{}^\complement \times अ^\complement_c =$$

$$\begin{bmatrix} 0-it_i & 0+ix_i & 0 & 0 \\ 0+iy_i & 0+iz_i & 0 & 0 \\ 0 & 0 & 0+ia_i & 0+ib_i \\ 0 & 0 & 0+ic_i & 0+id_i \end{bmatrix} \times \begin{bmatrix} 0-2i & 0 & 0 & 0 \\ 0 & 0-2i & 0 & 0 \\ 0 & 0 & 0-i & 0 \\ 0 & 0 & 0-i & 0 \end{bmatrix} =$$

$$\begin{bmatrix} -2t_i + 0 & 2x_i + 0 & 0 & 0 \\ 2y_i + 0 & 2z_i + 0 & 0 & 0 \\ 0 & 0 & a_i + b_i + 0 & 0 \\ 0 & 0 & c_i + d_i + 0 & 0 \end{bmatrix},$$

and with the 'charge/form factors'

$a_i = r\overline{b}$, $b_i = b\overline{r}$ and $c_i = +1$, $d_i = -1$ (or vice versa)

$$\begin{bmatrix} -2t_i + 0 & 2x_i + 0 & 0 & 0 \\ 2y_i + 0 & 2z_i + 0 & 0 & 0 \\ 0 & 0 & r\overline{b} + b\overline{r} + 0 & 0 \\ 0 & 0 & +1 - 1 + 0 & 0 \end{bmatrix} =$$

(unstable due to time violation)

$$\begin{bmatrix} -t_i + 0 & x_i + 0 & 0 & 0 \\ y_i + 0 & z_i + 0 & 0 & 0 \\ 0 & 0 & r\overline{b} + 0 & 0 \\ 0 & 0 & +1 + 0 & 0 \end{bmatrix} + \begin{bmatrix} -t_i + 0 & x_i + 0 & 0 & 0 \\ y_i + 0 & z_i + 0 & 0 & 0 \\ 0 & 0 & b\overline{r} + 0 & 0 \\ 0 & 0 & -1 + 0 & 0 \end{bmatrix}.$$

Step 2:

The produced gluon pair $r\overline{b}$ and $b\overline{r}$ exists for a short period

Δt with $\Delta t \approx \hbar / \Delta E$.

During that period the gluons move

from (x_i, y_i, z_i) to $(x_i+\Delta x, y_i+\Delta y, z_i+\Delta z)$ with

$\Delta s = (\Delta x^2 + \Delta y^2 + \Delta z^2)^{\frac{1}{2}} = c\, \Delta t$.

So, if $\Delta x = \Delta y = \Delta z$

=>

$\Delta s = (3\Delta x^2)^{\frac{1}{2}} = 3^{\frac{1}{2}}\,\Delta x = c\,\Delta t$

$\Delta x = \dfrac{c\Delta t}{\sqrt{3}}$

=>

$$\begin{bmatrix} -t_i - \Delta t + 0 & x_i + \dfrac{c\Delta t}{\sqrt{3}} + 0 & 0 & 0 \\[2mm] y_i + \dfrac{c\Delta t}{\sqrt{3}} + 0 & z_i + \dfrac{c\Delta t}{\sqrt{3}} + 0 & 0 & 0 \\[2mm] 0 & 0 & r\bar{b} + 0 & 0 \\[1mm] 0 & 0 & +1 + 0 & 0 \end{bmatrix} +$$

$$\begin{bmatrix} -t_i - \Delta t + 0 & x_i + \dfrac{c\Delta t}{\sqrt{3}} + 0 & 0 & 0 \\[2mm] y_i + \dfrac{c\Delta t}{\sqrt{3}} + 0 & z_i + \dfrac{c\Delta t}{\sqrt{3}} + 0 & 0 & 0 \\[2mm] 0 & 0 & b\bar{r} + 0 & 0 \\[1mm] 0 & 0 & -1 + 0 & 0 \end{bmatrix} =$$

$$\begin{bmatrix} 2(-t_i - \Delta t) + 0 & 2(x_i + \dfrac{c\Delta t}{\sqrt{3}}) + 0 & 0 & 0 \\[2mm] 2(y_i + \dfrac{c\Delta t}{\sqrt{3}}) + 0 & 2(z_i + \dfrac{c\Delta t}{\sqrt{3}}) + 0 & 0 & 0 \\[2mm] 0 & 0 & r\bar{b} + b\bar{r} + 0 & 0 \\[1mm] 0 & 0 & +1 - 1 + 0 & 0 \end{bmatrix},$$

and after multiplication with the annihilation operator अ'$_A$

$$
\begin{bmatrix}
0 + \dfrac{i}{2} & 0 & 0 & 0 \\
0 & 0 + \dfrac{i}{2} & 0 & 0 \\
0 & 0 & 0 + i & 0 \\
0 & 0 & 0 & 0
\end{bmatrix}
$$

we finally get

$$
\begin{bmatrix}
0 + i(-t_i - \Delta t) & 0 + i\left(x_i + \dfrac{c\Delta t}{\sqrt{3}}\right) & 0 & 0 \\
0 + i\left(y_i + \dfrac{c\Delta t}{\sqrt{3}}\right) & 0 + i\left(z_i + \dfrac{c\Delta t}{\sqrt{3}}\right) & 0 & 0 \\
0 & 0 & 0 + i(r\bar{b} + b\bar{r}) & 0 \\
0 & 0 & 0 + i(+1 - 1) & 0
\end{bmatrix}.
$$

Step 3:

The gluon pair has an actual energy (kinetic and spin as well as color charge) of $\Delta E = E_{r\bar{b}} + E_{b\bar{r}} \approx \hbar/\Delta t$;

the sites of gluon pair production have potential energy of $+ i\,(E_{r\bar{b}} + E_{b\bar{r}})/2$ before, and $+ (E_{r\bar{b}} + E_{b\bar{r}})/2$ after the action

of the creation operator अ'$_c$;

at the time of annihilation it flips back from

$+ (E_{r\bar{b}} + E_{b\bar{r}})/2$ to $+ i\,(E_{r\bar{b}} + E_{b\bar{r}})/2$.

This can be modeled by two operators acting on a simple core matrix $q_j^{\,"} =$

$$\begin{bmatrix} 0-it_j & 0+ix_j & 0 & 0 \\ 0+iy_j & 0+iz_j & 0 & 0 \\ 0 & 0 & 0 & 0 \\ 0 & 0 & 0 & 0+\dfrac{i\hbar}{2\Delta t} \end{bmatrix}.$$

The modified creation operator also needs an extension; its second matrix will be

$$\begin{bmatrix} 0-i & 0 & 0 & 0 \\ 0 & 0-i & 0 & 0 \\ 0 & 0 & 0 & 0 \\ 0 & 0 & 0 & 0-i \end{bmatrix} = अ“_c$$

$$\Rightarrow q_j“ \times अ“_c =$$

$$\begin{bmatrix} 0-it_j & 0+ix_j & 0 & 0 \\ 0+iy_j & 0+iz_j & 0 & 0 \\ 0 & 0 & 0 & 0 \\ 0 & 0 & 0 & 0+\dfrac{i\hbar}{2\Delta t} \end{bmatrix} \times \begin{bmatrix} 0-i & 0 & 0 & 0 \\ 0 & 0-i & 0 & 0 \\ 0 & 0 & 0 & 0 \\ 0 & 0 & 0 & 0-i \end{bmatrix} =$$

$$\begin{bmatrix} -t_j+0 & x_j+0 & 0 & 0 \\ y_j+0 & z_j+0 & 0 & 0 \\ 0 & 0 & 0 & 0 \\ 0 & 0 & 0 & \dfrac{\hbar}{2\Delta t}+0 \end{bmatrix},$$

and for the annihilation: q_k" x अ"$_A$ =

$$
\begin{bmatrix}
-t_j + 0 & x_j + 0 & 0 & 0 \\
y_j + 0 & z_j + 0 & 0 & 0 \\
0 & 0 & 0 & 0 \\
0 & 0 & 0 & \dfrac{\hbar}{2\Delta t} + 0
\end{bmatrix}
\times
\begin{bmatrix}
0 + i & 0 & 0 & 0 \\
0 & 0 + i & 0 & 0 \\
0 & 0 & 0 & 0 \\
0 & 0 & 0 & 0 + i
\end{bmatrix}
=
$$

$$
\begin{bmatrix}
0 - it_j & 0 + ix_j & 0 & 0 \\
0 + iy_j & 0 + iz_j & 0 & 0 \\
0 & 0 & 0 & 0 \\
0 & 0 & 0 & 0 + \dfrac{i\hbar}{2\Delta t}
\end{bmatrix}.
$$

Therefore, it is concluded that the sites of gluon pair production and annihilation can be modeled as the sites of a complex 4 x 4 x 4 tensor with

$$
\begin{bmatrix}
0 - 2i & 0 & 0 & 0 \\
0 & 0 - 2i & 0 & 0 \\
0 & 0 & 0 - i & 0 \\
0 & 0 & 0 - i & 0
\end{bmatrix}
= अ'$_C$
$$

$$
\begin{bmatrix}
0 - i & 0 & 0 & 0 \\
0 & 0 - i & 0 & 0 \\
0 & 0 & 0 & 0 \\
0 & 0 & 0 & 0 - i
\end{bmatrix}
= अ"$_C$
$$

$$
\begin{bmatrix}
0 - it_j & 0 + ix_j & 0 & 0 \\
0 + iy_j & 0 + iz_j & 0 & 0 \\
0 & 0 & 0 & 0 \\
0 & 0 & 0 & 0 + \dfrac{i\hbar}{2\Delta t}
\end{bmatrix} = q_j{}^{\text{“}}
$$

$$\updownarrow$$

$$
\begin{bmatrix}
-t_j + 0 & x_j + 0 & 0 & 0 \\
y_j + 0 & z_j + 0 & 0 & 0 \\
0 & 0 & 0 & 0 \\
0 & 0 & 0 & \dfrac{\hbar}{2\Delta t} + 0
\end{bmatrix} = q_k{}^{\text{“}}
$$

$$\updownarrow$$

$$
\begin{bmatrix}
-t_j + 0 & x_j + 0 & 0 & 0 \\
y_j + 0 & z_j + 0 & 0 & 0 \\
0 & 0 & 0 & 0 \\
0 & 0 & 0 & \dfrac{-\hbar}{2\Delta t} + 0
\end{bmatrix} = q_l{}^{\text{“}}
$$

(due to the release / uptake of $\Delta E = E_{r\bar{b}} + E_{b\bar{r}} \approx \hbar/\Delta t$)

$$
\begin{bmatrix}
0 + i & 0 & 0 & 0 \\
0 & 0 + i & 0 & 0 \\
0 & 0 & 0 & 0 \\
0 & 0 & 0 & 0 + i
\end{bmatrix} = अ{}^{\text{“}}_A
$$

$$\begin{bmatrix} 0+\dfrac{i}{2} & 0 & 0 & 0 \\ 0 & 0+\dfrac{i}{2} & 0 & 0 \\ 0 & 0 & 0+i & 0 \\ 0 & 0 & 0 & 0 \end{bmatrix} = अ'_A$$

So, they do have two faces, one with अ'c, processing the imaginary values of Q_I, and one with अ'A, processing the real values of Q_R.

07. An Analogy with Chess

In order to illustrate the proposed Microvita tensor, we can use a modified chess board:

So, let this be our modified chess board:

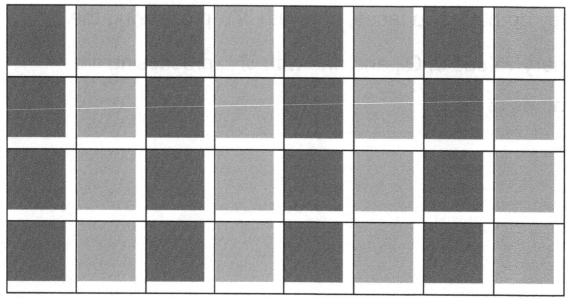

Figure 4

with

■ = real fields,

▩ = imaginary fields.

Then, let the number zero be symbolized by the pawns, the time coordinates by the knights, the space coordinates by the bishops, the 'charge/form factors' by the rocks, the energy factors by the queens, and the number one by the kings.

Consequently, the creation and annihilation operators will be as follows:

The outer creation operator अ'ᴄ

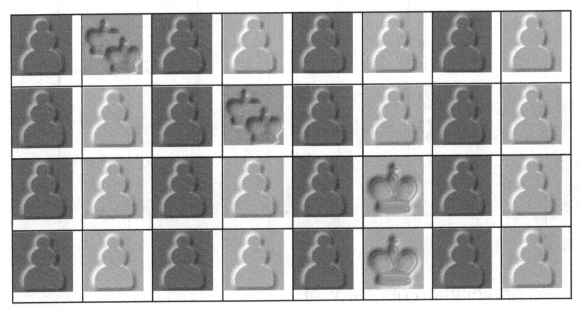

Figure 5

The inner creation operator अ"ᴄ

Figure 6

The inner annihilation operator अ"ₐ

Figure 7

The outer annihilation operator अ'ₐ

Figure 8

with

$$\text{[crown]} = +1i, \qquad \text{[crown]} = +i/2,$$

$$\text{[crown]} = -1i, \qquad \text{[crown]} = -2i.$$

In contrast, the core matrix will be in cyclical transmutation:

$q_j{}^{\text{"}}$

Figure 9

↓ (matrix multiplication with the inner creation operator)

39

$q_k{}''$

Figure 10

↓   (release of $\Delta E \approx \hbar / \Delta t$)

$q_i{}''$

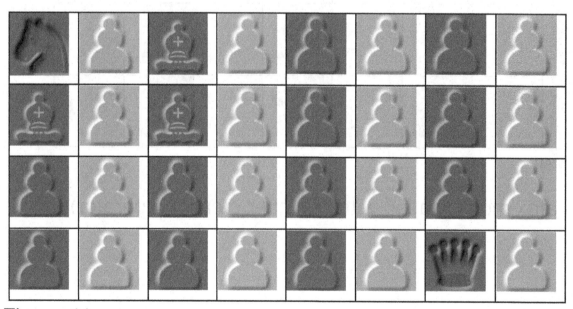

Figure 11

↓   (uptake of $\Delta E \approx \hbar / \Delta t$)

q$_k$"

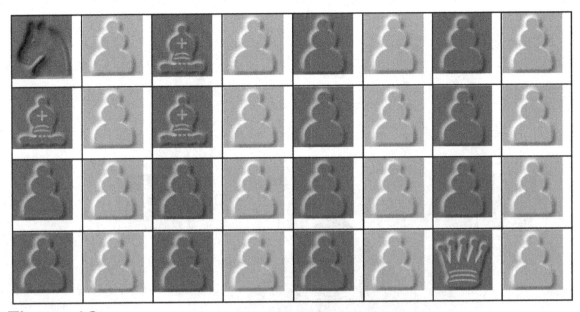

Figure 12

↓   (matrix multiplication with the inner annihilation operator)

q$_j$"

Figure 13

which is summarized in the following figure:

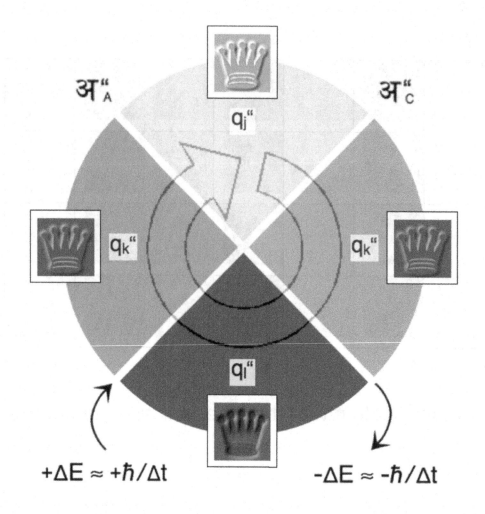

Figure 14: Transmutation of the Microvita core matrix with

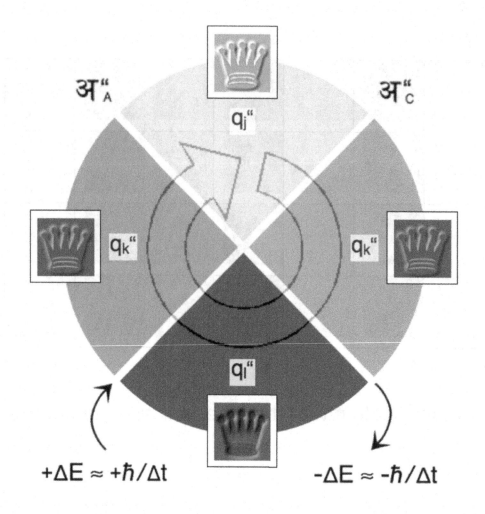 $= +i\,\hbar/2\Delta t,$

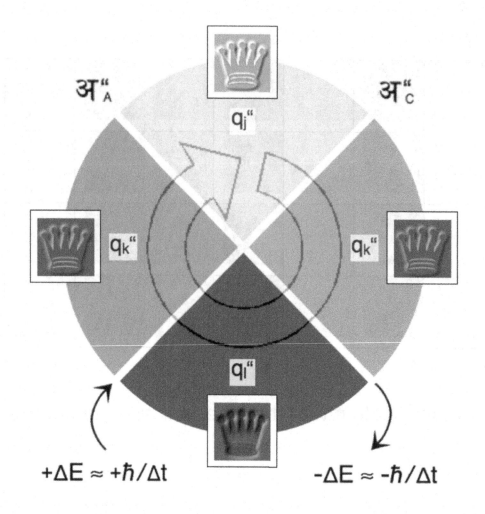 $= +\hbar/2\Delta t,$

 $= -\hbar/2\Delta t.$

08. Production of Electrons and Positrons

1. Positron-Electron Pair Production:

With the 'charge/form factors' $a_i = +e$, $b_i = -e$; $c_i = +\frac{1}{2}$, $d_i = -\frac{1}{2}$ (or vice versa), we get

$$q_i{}^{\prime} \times अ^{\prime}{}_c =$$

$$
\begin{bmatrix}
0-it_i & 0+ix_i & 0 & 0 \\
0+iy_i & 0+iz_i & 0 & 0 \\
0 & 0 & 0+ia_i & 0+ib_i \\
0 & 0 & 0+ic_i & 0+id_i
\end{bmatrix}
\times
\begin{bmatrix}
0-2i & 0 & 0 & 0 \\
0 & 0-2i & 0 & 0 \\
0 & 0 & 0-i & 0 \\
0 & 0 & 0-i & 0
\end{bmatrix}
=
$$

$$
\begin{bmatrix}
-2t_i+0 & 2x_i+0 & 0 & 0 \\
2y_i+0 & 2z_i+0 & 0 & 0 \\
0 & 0 & a_i+b_i+0 & 0 \\
0 & 0 & c_i+d_i+0 & 0
\end{bmatrix}
=
$$

$$
\begin{bmatrix}
-2t_i+0 & 2x_i+0 & 0 & 0 \\
2y_i+0 & 2z_i+0 & 0 & 0 \\
0 & 0 & +e-e+0 & 0 \\
0 & 0 & +\dfrac{1}{2}-\dfrac{1}{2}+0 & 0
\end{bmatrix}
=
$$

(unstable due to time violation)

$$\begin{bmatrix} -t_i+0 & x_i+0 & 0 & 0 \\ y_i+0 & z_i+0 & 0 & 0 \\ 0 & 0 & +e+0 & 0 \\ 0 & 0 & +\frac{1}{2}+0 & 0 \end{bmatrix} + \begin{bmatrix} -t_i+0 & x_i+0 & 0 & 0 \\ y_i+0 & z_i+0 & 0 & 0 \\ 0 & 0 & -e+0 & 0 \\ 0 & 0 & -\frac{1}{2}+0 & 0 \end{bmatrix}.$$

Which can be interpreted as a positron and an electron with electric charges +e and -e as well as spin +½ and -½, being created by the action of अ࣪c for a short period Δt with

$$\Delta t \approx \hbar / 2\,\Delta E = \hbar / 1022 \text{ keV.}$$

If, however, one or two photons supply an energy of $\Delta E > 1022$ keV (pumping), the lifetime of the produced electron-positron pair is no longer restricted by the energy-time uncertainty principle.

2. Leptogenesis:

Leptogenesis is the generic term for hypothetical processes producing an asymmetry between leptons and anti-leptons in the very early universe, resulting in a dominance of leptons over anti-leptons. This can be achieved by a modification of the creation operator अ࣪c, so that

प॑ x अ࣪c+ =

$$\begin{bmatrix} 0-it_i & 0+ix_i & 0 & 0 \\ 0+iy_i & 0+iz_i & 0 & 0 \\ 0 & 0 & 0+ia_i & 0+ib_i \\ 0 & 0 & 0+ic_i & 0+id_i \end{bmatrix} \times \begin{bmatrix} 0-i & 0 & 0 & 0 \\ 0 & 0-i & 0 & 0 \\ 0 & 0 & 0-i & 0 \\ 0 & 0 & 0 & 0 \end{bmatrix} =$$

$$\begin{bmatrix} -t_i+0 & x_i+0 & 0 & 0 \\ y_i+0 & z_i+0 & 0 & 0 \\ 0 & 0 & a_i+0 & 0 \\ 0 & 0 & c_i+0 & 0 \end{bmatrix} = \begin{bmatrix} -t_i+0 & x_i+0 & 0 & 0 \\ y_i+0 & z_i+0 & 0 & 0 \\ 0 & 0 & +e+0 & 0 \\ 0 & 0 & +\dfrac{1}{2}+0 & 0 \end{bmatrix}$$

and

$$q_i{}^{\prime} \times \text{अ}_{c-}^{\,\varsigma} =$$

$$\begin{bmatrix} 0-it_i & 0+ix_i & 0 & 0 \\ 0+iy_i & 0+iz_i & 0 & 0 \\ 0 & 0 & 0+ia_i & 0+ib_i \\ 0 & 0 & 0+ic_i & 0+id_i \end{bmatrix} \times \begin{bmatrix} 0-i & 0 & 0 & 0 \\ 0 & 0-i & 0 & 0 \\ 0 & 0 & 0 & 0 \\ 0 & 0 & 0-i & 0 \end{bmatrix} =$$

$$\begin{bmatrix} -t_i+0 & x_i+0 & 0 & 0 \\ y_i+0 & z_i+0 & 0 & 0 \\ 0 & 0 & b_i+0 & 0 \\ 0 & 0 & d_i+0 & 0 \end{bmatrix} = \begin{bmatrix} -t_i+0 & x_i+0 & 0 & 0 \\ y_i+0 & z_i+0 & 0 & 0 \\ 0 & 0 & -e+0 & 0 \\ 0 & 0 & -\dfrac{1}{2}+0 & 0 \end{bmatrix},$$

provided that अ$_{c-}^{\,\varsigma}$ is slightly more active than अ$_{c+}^{\,\varsigma}$

(as already mentioned, the lifetime of the produced electron or positron is unrestricted by the energy-time uncertainty principle, if one or two photons supply an energy of $\Delta E > 511$ keV).

09. Production of Devayonis and Pretayonis

Instead of applying the proposed mechanism to other elementary particles, I rather take a big stride forward by adapting it to a different class of universals.

With the 'charge/form factors'

a_i = Dy (one of the Devayonis*):

Yk	(Yaks'a)	or
Pr	(Prakrtiliina)	or
Kn	(Kinnara)	or
Vd	(Vidya'dhara)	or
Gd	(Gandharva)	or
Vd	(Videhaliina)	or
Sd	(Siddha),	and

$b_i = \overline{Dy}$ = Py (one of the Pretayonis*):

$\overline{Yk}$	=	Ps	(Pisha'ca)	or
$\overline{Pr}$	=	Md	(Madhyakapa'la)	or
$\overline{Kn}$	=	Kb	(Kabandha)	or
$\overline{Vd}$	=	Dm	(Durmukha)	or
$\overline{Gd}$	=	Br	(Brahmadaetya)	or
$\overline{Vh}$	=	Ak	(A'kashiipreta)	or
$\overline{Sd}$	=	Mh	(Maha'kapala),	

and $c_i = +x$, $d_i = -x$ (or vice versa)

* see Chapter 12

=> $q_I^\prime$ x अ$_C^\prime$ =

$$\begin{bmatrix} 0-it_i & 0+ix_i & 0 & 0 \\ 0+iy_i & 0+iz_i & 0 & 0 \\ 0 & 0 & 0+ia_i & 0+ib_i \\ 0 & 0 & 0+ic_i & 0+id_i \end{bmatrix} \times \begin{bmatrix} 0-2i & 0 & 0 & 0 \\ 0 & 0-2i & 0 & 0 \\ 0 & 0 & 0-i & 0 \\ 0 & 0 & 0-i & 0 \end{bmatrix} =$$

$$\begin{bmatrix} -2t_i+0 & 2x_i+0 & 0 & 0 \\ 2y_i+0 & 2z_i+0 & 0 & 0 \\ 0 & 0 & a_i+b_i+0 & 0 \\ 0 & 0 & c_i+d_i+0 & 0 \end{bmatrix} =$$

$$\begin{bmatrix} -2t_i+0 & 2x_i+0 & 0 & 0 \\ 2y_i+0 & 2z_i+0 & 0 & 0 \\ 0 & 0 & Dy+\overline{Dy}+0 & 0 \\ 0 & 0 & +x-x+0 & 0 \end{bmatrix} =$$

(unstable due to time violation)

$$\begin{bmatrix} -t_i+0 & x_i+0 & 0 & 0 \\ y_i+0 & z_i+0 & 0 & 0 \\ 0 & 0 & Dy+0 & 0 \\ 0 & 0 & +x+0 & 0 \end{bmatrix} + \begin{bmatrix} -t_i+0 & x_i+0 & 0 & 0 \\ y_i+0 & z_i+0 & 0 & 0 \\ 0 & 0 & \overline{Dy}+0 & 0 \\ 0 & 0 & -x+0 & 0 \end{bmatrix},$$

$q_I^\prime$ x अ$_{C+}^\prime$ =

$$
\begin{bmatrix}
0 - it_i & 0 + ix_i & 0 & 0 \\
0 + iy_i & 0 + iz_i & 0 & 0 \\
0 & 0 & 0 + ia_i & 0 + ib_i \\
0 & 0 & 0 + ic_i & 0 + id_i
\end{bmatrix}
\times
\begin{bmatrix}
0 - i & 0 & 0 & 0 \\
0 & 0 - i & 0 & 0 \\
0 & 0 & 0 - i & 0 \\
0 & 0 & 0 & 0
\end{bmatrix}
=
$$

$$
\begin{bmatrix}
-t_i + 0 & x_i + 0 & 0 & 0 \\
y_i + 0 & z_i + 0 & 0 & 0 \\
0 & 0 & a_i + 0 & 0 \\
0 & 0 & c_i + 0 & 0
\end{bmatrix}
=
\begin{bmatrix}
-t_i + 0 & x_i + 0 & 0 & 0 \\
y_i + 0 & z_i + 0 & 0 & 0 \\
0 & 0 & Dy + 0 & 0 \\
0 & 0 & +x + 0 & 0
\end{bmatrix},
$$

and

$$
q_i{}^{`} \times \text{अ}^{`}_{c\text{-}} =
$$

$$
\begin{bmatrix}
0 - it_i & 0 + ix_i & 0 & 0 \\
0 + iy_i & 0 + iz_i & 0 & 0 \\
0 & 0 & 0 + ia_i & 0 + ib_i \\
0 & 0 & 0 + ic_i & 0 + id_i
\end{bmatrix}
\times
\begin{bmatrix}
0 - i & 0 & 0 & 0 \\
0 & 0 - i & 0 & 0 \\
0 & 0 & 0 & 0 \\
0 & 0 & 0 - i & 0
\end{bmatrix}
=
$$

$$
\begin{bmatrix}
-t_i + 0 & x_i + 0 & 0 & 0 \\
y_i + 0 & z_i + 0 & 0 & 0 \\
0 & 0 & b_i + 0 & 0 \\
0 & 0 & d_i + 0 & 0
\end{bmatrix}
=
\begin{bmatrix}
-t_i + 0 & x_i + 0 & 0 & 0 \\
y_i + 0 & z_i + 0 & 0 & 0 \\
0 & 0 & \overline{Dy} + 0 & 0 \\
0 & 0 & -x + 0 & 0
\end{bmatrix}
=
$$

$$
\begin{bmatrix}
-t_i + 0 & x_i + 0 & 0 & 0 \\
y_i + 0 & z_i + 0 & 0 & 0 \\
0 & 0 & Py + 0 & 0 \\
0 & 0 & -x + 0 & 0
\end{bmatrix}.
$$

Which means that Devayonis and Pretayonis with spin +x or -x can be produced pairwise or apart by the action of the creation ($\text{अ}^{\prime}_{c}$) and the positive ($\text{अ}^{\prime}_{c+}$) or negative ($\text{अ}^{\prime}_{c-}$) creation operator respectively.

10. Interconnectedness of Microvita

A single Microvitum is, of course, quite insignificant; Microvita are effective, because they are able to form large networks. This begs the question of how the proposed model can be adjusted to allow such interconnectedness? To start with, I will highlight four aspects of Microvita network formation: Clustering, representation, synchronisation and coherency.

1. Clustering

Microvita need each other: The particles produced by an individual Microvitum will move away and won't be annihilated at the same site. So, there is a 'give and take' among Microvita, which brings them in spatial proximity. In a previous chapter they were assumed to occupy the nodes of a grid, in other cases they might have some degrees of positional freedom, allowing them to cluster like the particles in a fluid or a cloud.

2. Representation

Microvita do interact among themselves - but up to now there was no reason to expect this to be more than a chaotic process; in order to build up temporal patterns, each Microvitum will have to develop a sensitivity about its neighborer's activity. In a simple version, this can be achieved by installing a representation of the change in a neighborer's energy state ($\Delta E_N / \Delta t$).

3. Synchronisation

The core matrix of a Microvitum has been described as

$$\begin{bmatrix} 0 - it_j & 0 + ix_j & 0 & 0 \\ 0 + iy_j & 0 + iz_j & 0 & 0 \\ 0 & 0 & 0 & 0 \\ 0 & 0 & 0 & 0 + \dfrac{i\hbar}{2\Delta t} \end{bmatrix} = q_j{}^{"}$$

$$\updownarrow$$

$$\begin{bmatrix} -t_j + 0 & x_j + 0 & 0 & 0 \\ y_j + 0 & z_j + 0 & 0 & 0 \\ 0 & 0 & 0 & 0 \\ 0 & 0 & 0 & \dfrac{\hbar}{2\Delta t} + 0 \end{bmatrix} = q_k{}^{"}$$

$$\updownarrow$$

$$\begin{bmatrix} -t_j + 0 & x_j + 0 & 0 & 0 \\ y_j + 0 & z_j + 0 & 0 & 0 \\ 0 & 0 & 0 & 0 \\ 0 & 0 & 0 & \dfrac{-\hbar}{2\Delta t} + 0 \end{bmatrix} = q_l{}^{"}$$

As a block matrix, $q_j{}^{"}$ is composed of the four complex 2 x 2 matrices p_{11}, p_{12}, p_{21}, p_{22} with p_{12} and p_{21} being set to zero. If, however, one of their elements $a_{mn} + ib_{mn}$ is $\neq 0$, the product $q_j{}^{"}$ x अ$_c^{"}$ becomes messy.

Specifically with the elements $a_{13} + ib_{13}$ and $a_{23} + ib_{23}$ of $q_j{}^{"}$ being $\neq 0$, the space-time coordinates t_j and y_j get distorted, so that particle production becomes impossible. Nevertheless, with these elements changing back to zero, the equation turns operative, and particles can be produced as described in the previous chapters. This peculiarity can be used by putting $a_{13} + ib_{13}$ at $0 + i\Delta t/\Delta E_N$, which becomes zero at the time of a neighborer's particle production only. In this way, a cluster of Microvita should be able to synchronize its internal activity.

4. Coherency
In order to get coherency, two additional requirements are to be met:

a) the produced particles must have the same energy and
b) they must be unidirectional.

In analogy to what has been said before, the particles' energy becomes homogeneous, by setting $a_{23} + ib_{23}$ at $0 + i(E-E_N)$, which is zero only, if the adjacent Microvitum produces particles of the same energy.

For example, in order to describe photon pair production, I wrote

$p_I \times \text{अ}_c =$

$$\begin{bmatrix} +\lambda & -\lambda & 0 \\ +1 & -1 & 0 \end{bmatrix} = \begin{bmatrix} +\lambda & 0 \\ +1 & 0 \end{bmatrix} + \begin{bmatrix} -\lambda & 0 \\ -1 & 0 \end{bmatrix}$$

means

$q_I{}^{'} \times \text{अ}{}^{'}_c =$

$$
\begin{bmatrix}
0-it_i & 0+ix_i & 0 & 0 \\
0+iy_i & 0+iz_i & 0 & 0 \\
0 & 0 & 0+ia_i & 0+ib_i \\
0 & 0 & 0+ic_i & 0+id_i
\end{bmatrix}
\times
\begin{bmatrix}
0-2i & 0 & 0 & 0 \\
0 & 0-2i & 0 & 0 \\
0 & 0 & 0-i & 0 \\
0 & 0 & 0-i & 0
\end{bmatrix}
=
$$

$$
\begin{bmatrix}
-2t_i+0 & 2x_i+0 & 0 & 0 \\
2y_i+0 & 2z_i+0 & 0 & 0 \\
0 & 0 & a_i+b_i+0 & 0 \\
0 & 0 & c_i+d_i+0 & 0
\end{bmatrix}
=
$$

with $a_i = +\lambda$, $b_i = -\lambda$ and $c_i = +1$, $d_i = -1$ (or vice versa)

$$
\begin{bmatrix}
-2t_i+0 & 2x_i+0 & 0 & 0 \\
2y_i+0 & 2z_i+0 & 0 & 0 \\
0 & 0 & +\lambda-\lambda+0 & 0 \\
0 & 0 & +1-1+0 & 0
\end{bmatrix}
=
$$

(unstable due to time violation)

$$
\begin{bmatrix}
-t_i+0 & x_i+0 & 0 & 0 \\
y_i+0 & z_i+0 & 0 & 0 \\
0 & 0 & +\lambda+0 & 0 \\
0 & 0 & +1+0 & 0
\end{bmatrix}
+
\begin{bmatrix}
-t_i+0 & x_i+0 & 0 & 0 \\
y_i+0 & z_i+0 & 0 & 0 \\
0 & 0 & -\lambda+0 & 0 \\
0 & 0 & -1+0 & 0
\end{bmatrix}
$$

and with a modified creation operator अ"©

q_i" x अ"© =

$$
\begin{bmatrix}
0-it_j & 0+ix_j & 0+i\dfrac{\Delta t}{\Delta E_N} & 0 \\
0+iy_j & 0+iz_j & 0+i(E-E_N) & 0 \\
0 & 0 & 0 & 0 \\
0 & 0 & 0 & 0+i\dfrac{E}{2}
\end{bmatrix} \times
$$

$$
\begin{bmatrix}
0-i & 0 & 0 & 0 \\
0 & 0-i & 0 & 0 \\
0-i & 0 & 0 & 0 \\
0 & 0 & 0 & 0-i
\end{bmatrix} =
$$

$$
\begin{bmatrix}
-t_j+0 & x_j+0 & 0 & 0 \\
y_j+0 & z_j+0 & 0 & 0 \\
0 & 0 & 0 & 0 \\
0 & 0 & 0 & \dfrac{E}{2}+0
\end{bmatrix} = q_k\text{``}
$$

only if $i\dfrac{\Delta t}{\Delta E_N}=0$ and $i(E-E_N)=0$ (i.e. $E=E_N$).

Last but not least unidirectionality:
This feature cannot be brought about on the level of force particles production. It rather needs an environment allowing internal reflection. Our workshops in 2006, 2008 and 2009 provided insights into such settings, based on biological structures called microtubules. Details will be delivered in a subsequent paper.

11. Microvita in the Life of a Sadhaka
in collaboration with Michael Towsey, PhD

Traditionally, the basis of sadhana is introduced to an aspirant in terms of a mysterious life force called prana. This prana is supposed to circulate in the human body through countless micro-channels called nadis. The main nadis, however, are only three: the ida, pingala and sushumna. Although they are not physical, they are traditionally associated with the spinal cord. The ida starts at the base of the spine and ends at the left nostril. Flow through the ida is upwards and has a pro-psychic tendency. The pingala starts at the right nostril and ends at the base of the spine. Flow through the pingala is downwards and has a pro-physical tendency. The sushumna is a central channel opened through continued spiritual practice. Where the ida and pingala intersect the sushumna, these points are the known chakras. The free flow of pranic energy through the sushumna is considered essential for self-realisation.

Within a healthy human being, there are equal numbers of positive and negative microvita, but they are not distributed uniformly. Rather we can imagine two gradients – a gradient of negative microvita having highest concentration in the region of muladhara chakra and a gradient of positive microvita having highest concentration around the sahasrara chakra. Thus the ratio of positive to negative microvita differs throughout the body. While the natural tendency of positive microvita is to move upwards and of negative microvita to move downwards, nevertheless there is the possibility for both kinds of microvita to circulate around the entire human structure (the elevation of negative microvita into the psychic sphere takes place only under unusual circumstances and is detrimental to a person (13, p 60).

In the course of a normal active life, microvita are consumed or radiated from the body. The obvious example is that a person, engaged in physical activities, will use up negative microvita. The resultant deficiency will be made up by an inward flow of negative microvita, associated with crude tanmatras (i.e. in food, entertainments etc.). Psychic activity will consume more subtle microvita. And when a person does sadhana, very subtle positive microvita are somehow consumed or radiated from his or her 6th and 7th chakra. This will leave a temporary deficiency of positive microvita which can be replaced by an inflow of positive microvita associated with subtle tanmatras.

To summarise, the deficiencies of microvita arising as a result of daily life, can be replenished by an inflow of microvita travelling through the media of tanmatras. Human beings have some control over the tanmatras (and the associated microvita) which they take into their body. They can choose to tune their sense organs towards subtle or crude tanmatras in the same way as a television can be tuned to one channel/frequency or another. *"Good company, good books, good literature, good songs, kiirtans (devotional chanting) elevate the mind and the mind becomes sanctified*" (13, p 79). Conversely if the sense organs are directed to mundane pleasures, a preponderance of crude and negative microvita will become incorporated into the human body.

However, the human body is also subject to a continual flux of tanmatras and microvita, over which humans have little control. These might come from extra-terrestrial sources, in particular the sun and stars. The predominant extra-terrestrial tanmatras are sound (possibly in form of gravity) and light. The chapters 'Sidereal Year', 'Plexii and Microvita' and other sections in 'Microvita in a Nutshell' discuss this aspect of microvita science. Although humans cannot control the flux of such MV through their bodies and environment, they can have control over the microvita that will be incorporated into their physical and psychic body.

The physiology of sadhana (physico-psycho-spiritual practice) is best explained in terms of flows of microvita through the physical and psychic bodies of a person and the effect that these flows have on the person. The central concept is that flows of negative microvita, when they come in contact with the surrounding environment (in this case our human physical and psychic bodies) tend to increase the scope of the static portion and thus make the body cruder (13, p 79). Conversely, when positive microvita move through a human body they increase the scope of the sentient parts and the body becomes subtler. In particular, the process of sadhana increases the flow of positive microvita which makes the body more subtle.

Now, the question arises as to the balance of positive and negative microvita flowing into a person. We can distinguish five cases as shown in the table below. In the case of a normal person, the inflow of positive and negative microvita is in balance, but the flows are relatively low. This is the case of the better educated person, who tries to live a good life, but is not consciously pursuing a spiritual practice. The vitality of a person affects the volume of flow. Clearly an intellectually and physically active person will have greater flows of microvita than a lazy person. This is shown in Table 1.

Now consider the case of a person who is living only for mundane pleasure. There is a strong consumption of negative microvita, which are emitted through the lower chakras. This deficiency causes a flow of negative microvita into the body. Initially, this restores strength, but if it continues for long, this flow will cause rigidity and degeneration. Meanwhile, the upward flow of positive microvita ceases and the arena of the higher chakras becomes unimportant and stagnant.

Now consider the case of a highly dedicated musician, a subtle intellectual or a traditional indian sadhaka, who neglects the importance of negative microvita. In this case there is a strong consumption of positive microvita, but no corresponding flow of negative microvita to maintain the balance of their physical structure. In the long term such persons are in danger of becoming unstable and sick. Additionally, such people can become luminous bodies.

And now we come to the final case of a sadhaka of Vidya Tantra. The speciality of their sadhana is a high flow of positive and appropriate negative microvita through their human structure. The flow of positive microvita is accelerated by spiritual practice and the flow of negative microvita is encouraged by social service. Note in Figure 15 that there is also the possibility of a down-flow of positive microvita from the arena of higher chakras to chakras 4 and 5. This is to indicate that a highly developed sadhaka has the capacity to radiate positive microvita through speech and gestures, which ordinary people will recognise as charisma and charm. But it is important to note that a proper balance of positive and negative microvita has to be maintained. In fact, a delicately balanced but high-flow system of life is the very hallmark of Ananda Marga Tantra.

Type of persons	Flow of positive microvita	Flow of negative microvita
Normal person	normal	normal
Lazy person	low	low
Crude person	low	high
Subtle person	high	low
Vidya Tantra Sadhaka	high	high

Table 1

The relevant passage in 'Microvita in a Nutshell': *If there is over-utilisation of positive microvita during psychic and psycho-spiritual practices, then there will be a shortage of positive microvita for balancing the negative microvita in the physical and physico-psychic strata, because the sum total of positive and negative microvita must balance each other. Thus human beings should utilise their strength in all the strata. In the physical strata, utilise the immense power of negative microvita, otherwise the surplus of the negative microvita will become very strong due to accumulation. Good people should not go to the Himalayan caves. Rather, remaining in society, they should serve the society in the physical and physico-psychic strata with the help of negative microvita, and in the psychic and psycho-spiritual strata with the help of positive microvita. If good people neglect the use of negative microvita in the physical stratum, catastrophes are sure to come.* (p 154-155)

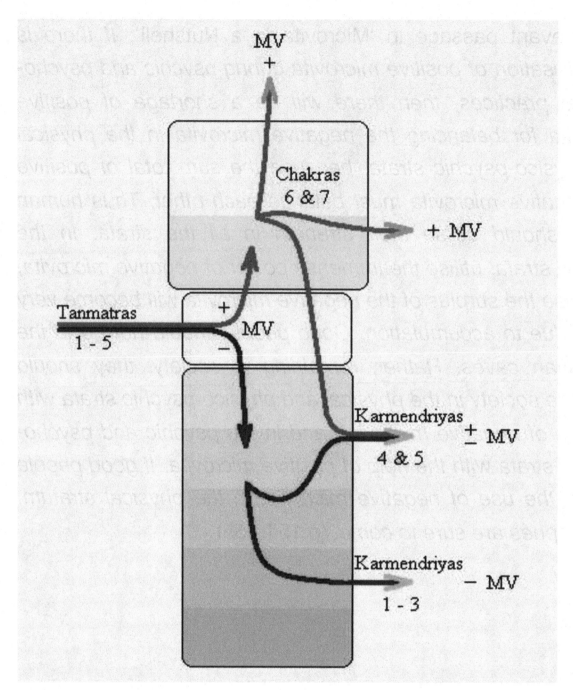

Figure 15: Schematic diagram of the major flows of positive and negative microvita through and within a human being.

12. Microvita in Social Life

Prabhat Rainjan Sarkar first breached the subject of Microvita in a Renaissance Universal talk on December 31st, 1986. Other discourses on the subject followed, until shortly before his death in October, 1990. They have been summarized in a book entitled 'MICROVITUM IN A NUTSHELL', comprising 37 chapters encompassing a huge variety of subjects (13).

Regarding social life, I can refer to Discourse No. 6 with the heading: 'NEO-ETHICS OF MULTILATERAL SALVATION'. The text has a length of four pages. I will dwell only on some quintessential points, the first of which is found at the end of the second page, where it says:

> When people get detached from non-carbonic pabula and become increasingly engrossed in carbonic pabula, there are two ill-effects as a consequence. First, the arena of one's own carbonic pabula will increase and the mind will gradually and steadily drift towards crude matter. Secondly, one's mind will think in terms of devouring other's carbonic pabula.

To understand this passage, it must first be clarified what is meant by 'carbonic' and 'non-carbonic pabula'. For this one must know that He associates carbon matter not only with 'carbonic', but with everything that is sustained by atoms; not only by carbon atoms, but by atoms in general. And He refers to everything that is not sustained by atoms as 'non-carbonic'. He takes the carbon atom as the representative of all atoms, because they dominate our life's reality; they form the skeleton of life; without them, life would not be possible in the form known to us. All of the other elements are, of course, significant, but they do not dominate as carbon atoms do. On the other hand, 'non-carbonic pabula' signifies all that is not sustained by atoms, but which is, nevertheless, of vital importance to us. It is known that a person is not only sustained by material things, he not only 'feeds' on matter, but he also needs the finer things in life: ideas, for example. Without ideas a person is aimless, incapable of action. However, the ideas that he needs to live are not made of atoms. Here, He says that if one's interest in such 'foods' gives way to an increased attraction to foods sustained by atoms, then certain ill-effects and consequences are bound to follow. One consequence is that it leads to a spiritual crudification; the emphasis on foods sustained by atoms leads to a crudification of the mind; and moreover, a tendency of our mind to seize and devour the atom-sustained foods of others, as well.

The crucial point here is that He not only says that the individual coarsens himself by such an imbalance, but that this coarseness affects his relations with others; and that he will then start to take away something from them. That which he devours is actually entitled to others, and he nevertheless takes it on himself, and develops a tendency to snatch more and more things from them.

He elaborates this further in the next thought:

> *This is the psychological explanation of imperialism. That is, imperialism has its origin in the psyche and functions in the psychic arena. When expressed externally it takes various forms, such as capitalism, state capitalism, communism, nationalism, etc.*

With this in mind, the motive that has just been addressed is pursued and placed in a bigger perspective. That is, He says that imperialism, together with its different forms, such as capitalism, state capitalism, communism and nationalism, has one psychological base, in which an increase of interest in carbonic foods usually goes hand in hand with a decrease in the demand for non-carbonic foods. To illustrate: we know not only the physiological hunger, but also the hunger for power or money and property; and accordingly, we speak of avarice and power hunger.

The other case deals with 'non-carbonic pabula', i.e. foods not based on atoms. One can also have hunger or a thirst for ideals. There are those, who throughout their lives, search for truth, beauty or justice. Here the thirst and the hunger after such values are put face to face with the hunger and thirst for material things. Persons with more interest in material things will not only coarsen, but will tend to take these things away from others – and in the extreme case, they will even rob them of what they need to their survival.

With these statements, the subject of Microvita acquires a sociological and political dimension.

In the next section He more closely characterizes His concept of imperialism, and initially cites:

> *Imperialism is a negative force. Such a poisonous radiation of black force attracts negative forces like negative Microvita. Those negative Microvita intensify and escalate the demonic activities of imperialism in all aspects of human society.*

In this sentence with few words He at first characterizes the negativity of imperialism and how adversely it affects individuals and society. Thus He uses expressions such as "...poisonous radiation of black force...." Consequently, He brings Microvita into the act. He says that this black strength draws negative Microvita, which leads to the fact that the demonic activities of imperialism are driven to their peaks. This interaction between imperialism and negative Microvita affects all aspects of human society. So, it is important to deal with the subject and to take suitable measures to weaken, avert and, in the ideal case, to prevent such processes as far as this is even possible.

What, however, does He mean by negative Microvita? To answer this question, it is helpful to look at Discourse No. 28 (13). This discourse is titled 'DISEMBODIED SOULS AND MICROVITA'. In it, distinction is made between so-called Pretayonis and Devayonis; and here He states that Devayonis belong to the positive, and Pretayonis to the negative Microvita. Subsequently, He divides each of the two groups into seven types (Table 2). To begin with, it is important for us to understand the qualities associated with the seven negative Microvita. He classifies the Pretayonis into seven different types, and names them:

> Pisha'ca,
>
> Madhyakapa'la,
>
> Kabandha,
>
> Durmukha,
>
> Brahmadaetya,
>
> A'kashiipreta,
>
> Maha'kapala.

With this concept He ties up with the Tantric tradition, and carries it further, interpreting it in a new context. First of all, He gives them a specific meaning:

He says that by Pisha'ca, everything entering our range of perception will be considered as an object of pleasure. Madhyakapa'la causes mental restlessness, the Kabandha a trend towards suicide (i.e. not necessarily that one kills oneself, but at least carrying this thought, or developing a feeling of wanting to be dead). Durmukha causes a tendency to involve others in difficulties, not so much in external ones, but rather into mental conflicts by trying to cause a psychic chain reaction in the people around, entrapping them into ever bigger psychic difficulties. Brahmadaetya is at the fifth place. It influences us to the effect so that we endeavor to suppress others, to use our spiritual force not for the good, but for the bad, or to rip them off after sounding their interests. Then at sixth place, there is A'ka'shiipreta. It drives a person to plan, commit and lead to an end a scheme, even a crime, for fulfilling own ambitions. A'ka'shiipreta is the one that has reached the negativity of crime. And, finally, Maha'kapala by which the crime increases so far, blowing up monstrously that not only own ambitions want to be fulfilled, but a tendency develops so as to destroy life on Earth.

If one reviews this list, one or the other example will occur to anyone quickly.

In December, 2005, we organized our first Microvita Workshop (Vig/Denmark), in which such examples were gathered and illustrated with citations, pictures and film clips (Table 3). The program covered more than 12 hours. Such an extensive material cannot be packed of course into this text, but I would still like to highlight two examples.

Workshop 2 centered on the influence of Pretayonis and Devayonis in historical events of the 20th century.

The Maha'kapala suggested itself to think of National Socialism. There was a film recently entitled 'Downfall' ('Der Untergang') in which the last days of the third empire are persuasively presented (19). In it there are two short sequences, where it becomes clear that Adolf Hitler was possessed by Maha'kapala:

In the first scene Hitler answers to the objections of a combat commander that there are still millions of civilians in Berlin and that they would first have to be evacuated:

Hitler: *I understand your concern, Mohnke, but we have to be cold as ice. We can take no consideration for so-called civilians.*

Mohnke: *My Führer, with all due respect, what should become of the women and children, the thousands of injured persons and the old people?*

Hitler: *In a war like this there are no civilians.*

It becomes even clearer in the second scene. At this point Albert Speer enters Hitler's room to discuss the further course of action with him.

Hitler: *Wherever the enemy penetrates, he should only find a desert.*

Speer: *This is the death sentence for the German people. No electricity, no gas, no clean water, no coal, no traffic, no railway equipment, canals, docks, sluices, ships, railroad engines, to destroy all this, will catapult our land back into the Middle Ages. With this command you will rob the people of any chance for survival.*

Hitler: *If the war becomes lost, the people will also be lost. It is not necessary to show consideration for the bases that the German people need to continue their most primitive lives. On the contrary, it is better to destroy even these things. Since the people have turned out to be the weaker, and the future exclusively belongs to the stronger eastern people.*

Speer: *These are your people, you are the leader.*

Hitler: *What's left after this struggle, are only the inferior, anyway, because the good ones have fallen.*

These extracts clearly express that a person obsessed by such Microvita wants to destroy not only his enemies, but also his people, his closest circle of advisers and himself. This accounts for the specialty of the Maha'kapala. They attempt a sweeping blow on all living beings.

However, as mentioned, there are still other Pretayonis. A'ka'shiipreta are not much less bad. They strive after the fulfillment of own ambitions with the help of crime. However, the crimes are committed not on one's own group, not on one's own body, but on others only. A good example of it are members of mafia organizations who want their own family to always be esteemed and sacred, and kept out under all circumstances. Their mercilessness always applies only to the other. This is the key difference between the A'ka'shiipreta and the Maha'kapala. There are examples galore of crime, also monstrous crimes, in the course of history.

As an example, we reported about the collaboration of American Secret Services with the American-Sicilian mafia, which in 1943/44 had led to the fact that the rule over Sicily was laid directly in their hands (20). This is only one example, and there are, of course, many other crimes that have been committed in the 20th century.

Now, however, back to the discourse from which we have come from, 'NEO-ETHICS OF MULTILATERAL SALVATION'.

After portraying the connection between negative Microvita and imperialism, He explains how we can master these problems. He writes:

> *In order to wipe out imperialism, which is rooted deep in the humanly psyche, Neo-ethics based on the following points are indispensable:*
>
> 1) *The Macrocosmic Entity must be accepted as the supreme desideratum in humanly life.*
>
> 2) *There should be a happy adjustment and balanced blending between carbonic and non-carbonic pabula.*

So it is that it requires neo-ethics, and that imperialism can only be overcome when this neo-ethics gains validity; and this new ethics must be based on the following principles. First, a macrocosmic unit must be recognized as the ultimate purpose of human life, and secondly, one must succeed in finding a happy blending of foods based on atoms and foods not based on atoms. The person does not live on bread alone, he also lives on sublime values, and both should stand in a reasonable relation with each other.

With these sentences the direction is set; how it will then be able to succeed will be subsequently outlined in detail. He thus says at the end of the discourse:

70

While following the path of Neo-ethics, human beings will attain enormous progress in the spiritual sphere with greater speed. With the help of non-carbonic pabula they will sharpen their psychic penetration within inter- and intra-atomic and molecular space. With the help of Microvita they will be able to powder down their ectoplasms and transform them into cognitive faculto-factors.

These sentences reveal that the negative forces can be overcome by the right application of positive Microvita.

In the discourse 'DISEMBODIED SOUL AND MICROVITA', which has already been quoted, besides the negative, also seven positive Microvita are listed. They are called:

> Yaks'a,
>
> Prakrtiliina,
>
> Kinnara,
>
> Vidya'dhara,
>
> Gandharva,
>
> Videhaliina,
>
> Siddha.

Yaks'a strengthens in a person the interest in collecting and accumulating money, originally for a good purpose. Then Prakrtiliina; it strengthens the tendency to sacrifice to effigies of the divine (e.g. to Fortuna, the goddess of luck). Kinnara strengthens the interest in physical beauty, Vidya'dhara that in mental beauty. Gandharva strengthens the interest in music.

Videhaliina causes a tendency to turn away from all worldly things, and the Siddha bestows occult forces to the aspirant.

Regarding the Siddha, Prabhat Rainjan Sarkar says that it is possible to penetrate into the inter- and intra-atomic and molecular space with the help of non-carbonic pabula, i.e. that it is possible to see spiritually what happens inside of an atom or a molecule. This ability is bestowed by positive Microvita, indeed not by any one, but by a certain Siddha. The various Siddhis are listed elsewhere (21); they comprise eight occult forces (Table 4), whereof the second is called Laghima. The Laghima Siddhi enables the follower "... *to enter the intra-molecular spaces of any entity by concentrating one's mental faculties, and directing that entity according to one's desire.*" It is of this Siddhi which enables us to influence others and other things from the inside. Hence in the given context, it is required for overcoming the psychic causes of imperialism.

In addition – and this is the last thought – positive Microvita enable us to refine our own thinking so that it becomes One with the general cognition. In this state a difference still exists between the own and the other, the knower and the known, the subject and the object; however, one can no longer be led by particular interests, but by a desire of service to the Whole.

Disembodied Souls

Devayonis	Related to	Pretayonis	Related to
Siddha	Attainment of occult powers		
Videhaliina	Rejection of worldly affairs		
Gandharva	Interest in music		
Vidya'dhara	Interest in mental beauty		
Kinnara	Interest in physical beauty		
Prakrtiliina	Worship of idols		
Yaks'a	Interest in collecting money	Pisha'ca	Tendency to take everything as an object of enjoyment
		Madhyakapa'la	Tendency to develop mental restlessness
		Kabandha	Tendency to commit suicide
		Durmukha	Tendency to give mental clash
		Brahmadaetya	Tendency to use intellect for the suppression of others
		A'ka'shiipreta	Tendency to commit crimes for the fulfillment of ambitions
		Maha'kapala	Tendency to destroy the living world

Table 2

INTERNATIONAL MICROVITA STUDY & RESEARCH WORKSHOP

| | | Face to Face with Microvita of All Kinds | | |
| | | Vig (Denmark) - 2005 | | |
DATE	TIME	Topic	Speaker	
Dec 29th				
	08:30 - 09:00	Keynote Address & Formal Inaugration	Acarya Vimalananda Avt	
	09:00 - 10:00	Basic Concepts of MV	Dr. Rudolph	
	10:15 - 12:30	WORKSHOP 1	Dr. Rudolph	
	15:00 - 16:00	Microvita in Micro- and Macrocosmos	Dr. Rudolph	
	16:15 - 18:00	WORKSHOP 2	Dr. Rudolph	
	20:00 - 21:00	Q & A	Acarya Vimalananda Avt	
Dec 30th				
	09:00 - 10:00	Face to Face with Negative Microvita	Acarya Vimalananda Avt	
	10:15 - 11:15	Face to Face with Positive Microvita	Acarya Vimalananda Avt	
	11:30 - 12:30	Discussion	Moderated by Dr. Rudolph	
	15:00 - 17:00	Closing Session	Co-ordinated by Acarya Vimalananda Avt	

Table 3

Aeshii Siddhi

Sanskrit	Description	Abilities	Warnings
An'ima'	Reducing one's mental arena to smallness and thereby being able to enter within and to know everything about an entity	To perform hypnotism, telepathy or clairvoyance	Might crudify
Laghima'	Being able to enter the intra-molecular spaces of any entity by concentrating one's mental faculties and directing that entity according to one's desire	To rise and 'fly in the skies'	Might madden
Mahima' (Garima')	Unfolding the inherent ubiquitous force of latent human qualities. Not attainable in Avidya' Tantra	To make oneself vast	The 'many-thorned Siddhi'
Iishitva	Engaging one's supra-mental and spiritual force for the welfare of the outside world	To rule	Might block an advancement
Vashitva	Engaging another person in good or bad actions according to one's own desire	To control others	More deadly than Iishitva-Siddhi
Pra'pti	Visualising any factor of the psychic world within oneself	To be omniscient	A thorny Siddhi
Praka'mya	Being successful by forgetting both the physical and psychic worlds and flowing one's mind only towards Parama Purus'a	Getting what is desired	Has to be applied cautiously
Antarya'mitva	Filling another's with one's own mind-stuff and thereby knowing every iota of his or her ectoplasmic vibration	To be omniscient	

Table 4

INTERNATIONAL MICROVITA STUDY & RESEARCH WORKSHOP

Face to Face with Microvita of All Kinds

Vig (Denmark) - 2006

DATE	TIME	Topic	Speaker
Dec 28th			
	08:30 - 09:00	Keynote Address & Formal Inauguration	Acarya Vimalananda Avt
	09:00 - 10:00	Face to Face with Microvita of All Kinds	Acarya Vimalananda Avt
	10:15 - 12:30	Workshop 1	Acarya Vimalananda Avt
	15:00 - 16:00	Basic Concepts	Dr. Rudolph
	16:15 - 18:00	Workshop 2	Dr. Rudolph, K.-B. Sidhu
	20:00 - 22:00	Spiritual Programme	
Dec 29th			
	09:00 - 10:00	Microvita in Biopsychology (Part I)	Dr. Rudolph
	10:15 - 12:30	Workshop 3	Dr. Rudolph, Acarya Vimalananda Avt
	15:00 - 16:00	Microvita in Death and Rebirth	Dr. Rudolph
	16:15 - 18:00	Workshop 4	Dr. Rudolph, Acarya Vimalananda Avt
	20:00 - 22:00	Spiritual Programme	
Dec 30th			
	09:00 - 10:00	Microvita in Biopsychology	Acarya Vimalananda Avt
	10:15 - 12:30	Panel Discussion	All Speakers
	15:00 - 16:45	Organising MRI	Acarya Vimalananda Avt, Dr. Rudolph
	17:00 - 18:00	Closing Session	Co-ordinated by Ac Vimalananda Avt

Table 5

3rd INTERNATIONAL MICROVITA STUDY & RESEARCH WORKSHOP

Face to Face with Microvita of All Kinds
Berlin - 2008

DATE	TIME	Topic	Speaker
May 2nd			
	09:00 - 09:30	Keynote address & formal inauguration	Acarya Vimalananda Avt.
	09:30 - 11:00	Microvita and tanmatras	Dr. Rudolph
	11:15 - 12:00	String theory in a nutshell	K.-B. Sidhu
	12:00 - 12:30	Discussion	
	12:30 - 14:30	Lunch break	
	14:30 - 16:00	Microvita and elementary particles	Dr. Rudolph
	16:00 - 16:30	Snack	
	16:30 - 17:15	Superluminal quantum model of the electron and photon	Dr. Gauthier
	17:15 - 18:00	Discussion	
	18:15 - 19:00	Collective meditation	
	19:00 - 20:30	Dinner	
	20:30 - 22:00	Microvita and theatre arts	O. Brekke

3rd INTERNATIONAL MICROVITA STUDY & RESEARCH WORKSHOP

Face to Face with Microvita of All Kinds
Berlin - 2008

DATE	TIME	Topic	Speaker
May 3rd			
	06:30 - 07:15	Collective meditation	
	07:30 - 08:15	Breakfast	
	08:30 - 09:30	Microvita in biological life	Dr. Rudolph
	09:30 - 10:15	Discussion	
	10:30 - 11:30	Microvita in the life of a sadhaka	Dr. Rudolph
	11:30 - 12:30	Discussion	
	12:30 - 14:30	Lunch break	
	14:30 - 15:30	Microvita and theatre arts	O. Brekke
	15:30 - 16:00	Snack	
	16:00 - 17:00	Microvita and theatre arts	O. Brekke
	17:15 - 18:00	The 'how-to' of dealing with positive and negative Microvita	Acarya Vimalananda Avt.
	18:15 - 19:00	Collective meditation	
	19:00 - 20:30	Dinner	
	20:30 - 22:00	Excursion	

3rd INTERNATIONAL MICROVITA STUDY & RESEARCH WORKSHOP

Face to Face with Microvita of All Kinds
Berlin - 2008

DATE	TIME	Topic	Speaker
May 4th			
	06:30 - 07:15	Collective meditation	
	07:30 - 08:15	Breakfast	
	08:30 - 11:30	Workshop	
	11:45 - 12:30	Precursors of Microvita concept in western thought	Dr. Ragbourn
	12:30 - 14:30	Lunch break	
	14:30 - 16:00	Musical theory and practice	M. Hackenberger
	16:00 - 16:30	Snack	
	16:30 - 17:15	Panel discussion	
	17:15 - 18:00	In conclusion - where to next ...	Acarya Vimalananda Avt.

Table 6

INTERNATIONAL MICROVITA STUDY AND RESEARCH WORKSHOP

(Face to Face with Microvita of All Kinds)

Berlin, May 01- 03 2009.

DATE	TIME	TOPIC	SPEAKER
May 1st	09:00 - 09:30	Key Note Address & Formal Introductory	Dr. Rudolph
	09:30 - 11:00	Microvita -- The Mysterious Emanations	Acarya Vimalananda Avt.
	11:00 - 11:15	Break	
	11:15 - 12:30	Divinity in Action	Dr. Ragbourn
	12:30 - 14:30	Individual Sadhana / Lunch / Maona Vrata	
	14:30 - 16:00	On the Nature of Particles	Dr. Williamson
	16:00 - 16:30	Snack Break	
	16:30 - 18:00	Microvita & the Origins of Matter	T. Tatsuyama-Kurk
	18:15 - 19:15	Collective Meditation	
	19:15 - 20:30	Dinner	
	20:30 - 21:30	Non-Locality in Quantum Physics	M. Drenscko
	22:00 - 23:00	Microvita & Sadhana I	Acarya Vimalananda Avt.

DATE	TIME	TOPIC	SPEAKER
May 2nd	09:00 - 09:30	Summary from Day 1	Dr. Rudolph
	09:30 - 11:00	Microvita Cosmology I	Acarya Vimalananda Avt.
	11:00 - 11:15	Break	
	11:15 - 12:30	The 4-Chambered Universe - a New Grammar	T. Tatsuyama-Kurk
	12:30 - 14:30	Individual Sadhana / Lunch / Maona Vrata	
	14:30 - 16:00	Workshop	Dr. Rudolph
	16:00 - 16:30	Snack Break	
	16:30 - 18:00	Microvita & Biopsychology I	Acarya Vimalananda Avt.
	18:15 - 19:15	Collective Meditation	
	19:15 - 20:30	Dinner	
	20:30 - 22:00	Microvita in Music	M. Hackenberger
	22:15 - 23:00	Microvita & Sadhana II	Acarya Vimalananda Avt.

INTERNATIONAL MICROVITA STUDY AND RESEARCH WORKSHOP

DATE	TIME	TOPIC	SPEAKER
		(Face to Face with Microvita of All Kinds)	
May 3rd	09:00 - 09:30	Summary from Day 2	Acarya Vimalananda Avt.
	09:30 - 11:00	Microvita & Biopsychology II	Dr. Rudolph
	11:00 - 11:15	Break	
	11:15 - 12:30	Creative Games	Acarya Krpamayananda Avt.
	12:30 - 14:30	Individual Sadhana / Lunch / Maona Vrata	
	14:30 - 16:00	Q & A Panel	Inclusive
	16:00 - 16:30	Snack Break	
	16:30 - 18:00	Closing and Departure…	Inclusive

Table 7

13. Microvita and the Neo-Leibnizian World Model

In the years 1712 to 1714 Gottfried Wilhelm Leibniz (1646-1716) wrote two texts, later composed to a booklet, comprising ninety paragraphs on simple substances, which he called monads (22). The position delivered therein can be seen in a metaphysical tradition, once started by the Pythagoreans in the sixth century BC. According to their understanding the monad is the first being, the One without division. By the eleventh century AC, the pythagorean ideas had consolidated to twentyfour aphorisms about God, known as the 'Book of the XXIV Philosophers' (23). In its second paragraph, God is described as an intelligible sphere, whose center is everywhere and circumference nowhere. Leibniz got acquainted with these thoughts via the works of Nicholas of Cusa (1401-1464 AC), whom he appreciated to such an extent that he utilized his notions and images, including the one of 'creatures mirroring each other as well as God' (24). With this in mind he could proceed from the pythagorean to his own concept of monads, being the real atoms of nature "and, in a word, the elements of things" (§ 3).

So far, so good - but in 1992, Teruaki Nakagomi coined the term 'quantum monadology' (25). It became the starting point for an ongoing effort to bring the antique esoteric ideas to a scientific level. Moreover, he used mathematics for what had previously been described in words only. Consequently, his approach could be computerized and tested, yielding nothing less but what he called the 'Quantum Neo-Leibnizian World Model'.

Ten years later, he gave the following outlines (26):

Let

$$W = (V,F,L,\eta,\rho,\omega,\lambda,\beta)$$

be the quantum NL world to be constructed. The set V of monad-images can be any finite set with an element specified as the self-image v_{self}. The main idea in defining the other items of W is that the internal world of a monad is described by a quantum state and Lorentz frames associated with the monad-images. This main idea and the purpose of the quantum NL world almost determine the structure of W.

The set V is decomposed into two subsets $\{v_{self}\}$ and $V_{other} = V - \{v_{self}\}$. The quantum state of a monad is also decomposed into the tensor product of the self part and the other part, and the self–other coupling of the quantum state defines the list of choices ω. The choices by monads appear as quantum reduction processes of self–other coupled states. The preferability ρ is related to the reduction probability. The choice-driven part of state-change operator β represents this reduction. The automatic part of β is defined so as to cause inhomogeneous change of Lorentz frames of monad-images. The interpreter λ is specified by frame–frame relation and self–other conversion of monad-images. Finally, the appetite η is given by the entropy of distribution of ρ over ω ... (additionally, F signifies the internal states and L the contents of consciousness ... where consciousness should not be given direct interpretation by our daily experiences, but as an elemental and primitive origin of those of the human level).

Furthermore, he presented 3 rules:

Rule 1 (monads and correspondences)
With W, a set M_W of monads is associated, whose cardinal number is the same as that of V. Each monad $i \in M_W$ has a one-to-one and onto correspondence $c_i: j \in M_W \rightarrow c_i j \in V$
that satisfies $c_i i = v_{self}$.

Rule 2 (current states)
Each monad $i \in M_W$ has a variable Φ_i taking values in F, called the current state of the monad i. The current state variable Φ_i of each monad $i \in W$ has three main components:

$\Phi_i = (\Theta_i, \Gamma_{i*}, \Psi_i),$

and Γ_{i*} has 2 $|V|$ subcomponents:

$\Gamma_{i*} = \{\Gamma_{iv}\}_{v \in V} = \{X_{iv}, \Xi_{iv}\}_{v \in V}.$

Rule 3 (Choice and renewal)
The current states of monads are renewed by iteration of the following process: Each monad decides to choose an element from $\omega(\Phi_m)$ or to do nothing.

Besides obvious differences, there are also striking similarities to the Microvita model presented in this book:

In both cases, the theory is based on algebras corresponding to complex Lie groups, i.e. the special linear group $SL(n, \mathbb{C})$ in Nakagomi's and the general linear group $GL(n, \mathbb{C})$ in my case. Therefore, the applied elements are square (2 x 2 or 4 x 4), invertible matrices over the field of complex numbers.

A difference is made by the determinants, with $SL(n,\mathbb{C}) = \{\xi \in M(n,\mathbb{C}) \mid \det\xi = 1\}$, and $GL(n,\mathbb{C}) = \{\xi \in M(n,\mathbb{C}) \mid \det\xi \neq 0\}$ respectively. Moreover, the determinants of the described core matrices as well as those of the creation and annihilation operators equal zero; hence, these matrices are singular, which means that they don't belong to any Lie group.

A Minkowski space is produced by Nakagomi's monads as well as by Microvita [4], although in different ways.

The tensor product is constitutional for both, monads and Microvita.

[4] Multiplication of the mutually orthogonal vectors -i_0, i_1, i_2 and i_3 with $अ_6 \in M_I$ provides the mutually orthogonal vectors -e_0, e_1, e_2 and e_3, which are constitutional for the real Minkowski space.

14. Microvita and the Grid

In 'Microvita and the Nucleon Model' Prabhat Rainjan Sarkar was quoted by saying: "... *Brahma, comprises two aspects: the cognitive and the operative aspect. These two aspects together constitute one entity. That was the state where there was no scope of expression. So in that state, ultimately, Parama Purus'a determined, 'I am one today, but let Me be many.' 'At the moment I am one, let Me be many.' When this Sam'kalpa was aroused in Its mind, the cosmic mind began.*" (27)

Therefrom I concluded that the fundamental lattice, having a site-to-site distance just above Planck length, is made of nothing but the multiplied cognitive and operative aspects of the supreme causal factor 'Brahma'.

Elsewhere, I assumed a lattice with a site-to-site distance of 2-3 millifermi (10^{-18} m) to exist inside nucleons; there, each site was supposed to produce and annihilate $1.6 - 5.2 \times 10^{+15}$ pairs of virtual gluons per second. [5]

And in 'The two Faces of Microvita', I presented a complex 4 x 4 x 4 tensor.

In other words, I assumed grids at three scales of magnitude, i.e. with site-to-site distances D_s in the range of 10^{-35}, 10^{-18} and $10^{-35} < D_s < 10^{-18}$ m. Their elements were

1. the 'cognitive and operative aspects of the supreme causal factor',

2. the numbers 0, -i, +i, -2i, +i/2, and

3. the sites of gluon pair production and annihilation being identified with nuclear Microvita.

Here, the question arises whether these postulates are fictitious or somehow scientifically grounded. To support the latter, I refer to a book written by the 2004 Nobel Laureate in Physics Frank Wilczek (28). In Chapter 8: The Grid (Persistence of Ether), he concludes:

· *The Grid fills space and time.*

· *Every fragment of Grid - each space and time element has the same basic propensities as every other fragment.*

· *The Grid is alive with quantum activity. Quantum activity has special characteristics. It is spontaneous and unpredictable. And to observe quantum activity, you must disturb it.*

· *The Grid also contains enduring, material components. The cosmos is a multilayered, multicolored superconductor.* [5]

· *The Grid contains a metric field that gives space-time rigidity and causes gravity.*

· *The Grid weighs, with a universal density.*

So, I dare to say that Frank Wilczek's 'metric field' can be identified with Prabhat Rainjan Sarkar's multiplied 'cognitive and operative aspects of the supreme causal factor'; and then the other two would be aspects of what Wilczek describes as a 'multilayered, multicolored super-conductor'.

[5] If the proposed nuclear lattice would provide superconductivity, the rate of gluon pair production and annihilation mentioned in Chapter 5 could decrease significantly.

15. Microvita and Non-Duality

As already stated, the imaginary precursors of force particles (q_i') combine opposites (ia_i, $i\bar{a}_i$, ic_i, $i\bar{c}_i$) into one matrix; they undergo polarization only after being transformed into real space - which means that multiplication with the described creation operator actually amounts to a process of symmetry breaking:

q_i' x अ'$_c$ =

$$\begin{bmatrix} 0-it_i & 0+ix_i & 0 & 0 \\ 0+iy_i & 0+iz_i & 0 & 0 \\ 0 & 0 & 0+ia_i & 0+i\bar{a}_i \\ 0 & 0 & 0+ic_i & 0+i\bar{c}_i \end{bmatrix} \times \begin{bmatrix} 0-2i & 0 & 0 & 0 \\ 0 & 0-2i & 0 & 0 \\ 0 & 0 & 0-i & 0 \\ 0 & 0 & 0-i & 0 \end{bmatrix} =$$

$$\begin{bmatrix} -t_i+0 & x_i+0 & 0 & 0 \\ y_i+0 & z_i+0 & 0 & 0 \\ 0 & 0 & a_i+0 & 0 \\ 0 & 0 & c_i+0 & 0 \end{bmatrix} + \begin{bmatrix} -t_i+0 & x_i+0 & 0 & 0 \\ y_i+0 & z_i+0 & 0 & 0 \\ 0 & 0 & \bar{a}_i+0 & 0 \\ 0 & 0 & \bar{c}_i+0 & 0 \end{bmatrix}.$$

In other words: Before transformation, the 'charge/form factors' ia_i, $i\bar{a}_i$, ic_i and $i\bar{c}_i$ exist as oxymora. Thereafter, they separate into antagonistic entities. And after multiplication with the described annihilation operator

$$\text{अ`}_A = \begin{bmatrix} 0 + \dfrac{i}{2} & 0 & 0 & 0 \\ 0 & 0 + \dfrac{i}{2} & 0 & 0 \\ 0 & 0 & 0 + i & 0 \\ 0 & 0 & 0 & 0 \end{bmatrix}$$

we get:

$$\begin{bmatrix} 0 - it_i & 0 + ix_i & 0 & 0 \\ 0 + iy_i & 0 + iz_i & 0 & 0 \\ 0 & 0 & 0 + i(a_i + \overline{a_i}) & 0 \\ 0 & 0 & 0 + i(c_i + \overline{c_i}) & 0 \end{bmatrix},$$

which means not only that two entities (matrices) are again merged into one, but also that the 'charge/form factors' are synthesized to the new terms $i(a_i + \overline{a_i})$ and $i(c_i + \overline{c_i})$.

With the spin and electric charge of bosons, leptons and fermions, however, $i(a_i + \overline{a_i})$ and $i(c_i + \overline{c_i})$ simply equal zero
(spin +1 and -1 or +½ and -½, electric charge +e and -e, +⅔ e and -⅔ e, +⅓ e and -⅓ e or 0 respectively).

So, in these cases

$$\begin{bmatrix} 0 - it_i & 0 + ix_i & 0 & 0 \\ 0 + iy_i & 0 + iz_i & 0 & 0 \\ 0 & 0 & 0 + i(a_i + \overline{a_i}) & 0 \\ 0 & 0 & 0 + i(c_i + \overline{c_i}) & 0 \end{bmatrix} = \begin{bmatrix} 0 - it_i & 0 + ix_i & 0 & 0 \\ 0 + iy_i & 0 + iz_i & 0 & 0 \\ 0 & 0 & 0 & 0 \\ 0 & 0 & 0 & 0 \end{bmatrix}.$$

With the gluons' and mesons' color charge, however, only $(r\bar{r} + b\bar{b} + g\bar{g})$ equals zero, whereas $r\bar{r}$, $b\bar{b}$ and $g\bar{g}$, as well as $(r\bar{r} - b\bar{b})$ and $(r\bar{r} + b\bar{b} - 2g\bar{g})$ equal nonzero numbers.

Likewise with Deva- and Pretayonis:

$$Dy + \overline{Dy} \neq 0.$$

Therefore, as long as antagonistic 'charge/form factors' $i(\overline{a_i + b_i})$ and/or $i(\overline{c_i + d_i})$ can be found, it should be possible to iterate the process, producing more and more complex forms, proceeding with

$$q_I{}' = \begin{bmatrix} 0 - it_i & 0 + ix_i & 0 & 0 \\ 0 + iy_i & 0 + iz_i & 0 & 0 \\ 0 & 0 & 0 + i(a_i + b_i) & 0 + i(\overline{a_i + b_i}) \\ 0 & 0 & 0 + i(c_i + d_i) & 0 + i(\overline{c_i + d_i}) \end{bmatrix}.$$

Hereunder, each cycle will be made up of three phases:

Imaginary duality => real duality => imaginary non-duality => imaginary duality => real duality => imaginary non-duality etc.*

In Sanskrit this cycle is called: advaetadvaeta'dvaetava'da, meaning non-dualistic dualistic non-dualism, which is described in the macro scenario as

"... non-dualistic in the beginning, residing in the heart of Parama Purus'a. Parama Purus'a was One, but He desired to become many in a sportive mood. Eko'ham' bahusya'm -- 'I was One but I became many.' Later, after playing with all, He finally called them unto Him, saying, 'My children, return to Me. The day is over, it is evening. Return home.' Not only is Vrajagopa'la the quintessence of all human sensibilities and the nucleus of the human mind, but His heart is the last and final shelter of all the jiivas of this universe." (29)

* With color charge, for example, we get:

1. cycle

$i\,a_i = i\,r,$

$i\,b_i = i\,\overline{r} \Rightarrow$

$a_i = r,$

$b_i = \overline{r} \Rightarrow$

$i(a_i + b_i) = i(r + \overline{r}) \Rightarrow$

2. cycle

$i(a_i + b_i) = i(r + \overline{r}),$

$i\overline{(a_i + b_i)} = i(b + \overline{b}) \Rightarrow$

$(a_i + b_i) = (r + \overline{r}),$

$\overline{(a_i + b_i)} = (b + \overline{b}) \Rightarrow$

$i[(a_i + b_i) + \overline{(a_i + b_i)}] = i[(r + \overline{r}) + (b + \overline{b})] \Rightarrow$

3. cycle

$i[(a_i + b_i) + \overline{(a_i + b_i)}] = i[(r + \overline{r}) + (b + \overline{b})],$

$i\overline{[(a_i + b_i) + \overline{(a_i + b_i)}]} = i[-(g + \overline{g}) - (g + \overline{g})] \Rightarrow$

$[(a_i + b_i) + \overline{(a_i + b_i)}] = [(r + \overline{r}) + (b + \overline{b})],$

$\overline{[(a_i + b_i) + \overline{(a_i + b_i)}]} = [-(g + \overline{g}) - (g + \overline{g})] \Rightarrow$

$i\{[(a_i + b_i) + \overline{(a_i + b_i)}] + \overline{[(a_i + b_i) + \overline{(a_i + b_i)}]}\} =$

$i[(r + \overline{r}) + (b + \overline{b}) - (g + \overline{g}) - (g + \overline{g})] =$

$i(r\overline{r} + b\overline{b} - 2g\overline{g}) \triangleq$ Gell - Mann matrix λ_8 (3).

16. Microvita and Quantum Entanglement

At our 2009 Microvita Study and Research Workshop, Mihaela Drescko from Hunter College, City University of New York gave a talk via video conference about non-locality in quantum physics. Since then, the question of how to align Microvita with quantum entanglement kept us on the run.

Research into this topic had actually started as early as 1935 with Einstein, Podolsky and Rosen's famous EPR paradox paper (31). At that time, they aimed to prove the incompleteness of quantum mechanics; nevertheless, entanglement got recognized as one of its fundamental features meanwhile.

The gist of it is that if you have, for example, a pair of entangled photons and you change the spin of one, the other changes its spin to the opposite direction; and this happens immediately, even if they are really far apart.

Now, here is an explaination in terms of the preceding reflections:

Regarding photon pair production, I wrote

$$q_I{}^\prime \times \text{अ}_c^{\prime} =$$

$$\begin{bmatrix} 0-it_i & 0+ix_i & 0 & 0 \\ 0+iy_i & 0+iz_i & 0 & 0 \\ 0 & 0 & 0+ia_i & 0+ib_i \\ 0 & 0 & 0+ic_i & 0+id_i \end{bmatrix} \times \begin{bmatrix} 0-2i & 0 & 0 & 0 \\ 0 & 0-2i & 0 & 0 \\ 0 & 0 & 0-i & 0 \\ 0 & 0 & 0-i & 0 \end{bmatrix} =$$

$$\begin{bmatrix} -2t_i + 0 & 2x_i + 0 & 0 & 0 \\ 2y_i + 0 & 2z_i + 0 & 0 & 0 \\ 0 & 0 & a_i + b_i + 0 & 0 \\ 0 & 0 & c_i + d_i + 0 & 0 \end{bmatrix} =$$

with $a_i = +\lambda$, $b_i = -\lambda$ (wavelength) and $c_i = +1$, $d_i = -1$ (spin)

$$\begin{bmatrix} -2t_i + 0 & 2x_i + 0 & 0 & 0 \\ 2y_i + 0 & 2z_i + 0 & 0 & 0 \\ 0 & 0 & +\lambda - \lambda + 0 & 0 \\ 0 & 0 & +1 - 1 + 0 & 0 \end{bmatrix} =$$

(unstable due to time violation)

$$\begin{bmatrix} -t_i + 0 & x_i + 0 & 0 & 0 \\ y_i + 0 & z_i + 0 & 0 & 0 \\ 0 & 0 & +\lambda + 0 & 0 \\ 0 & 0 & +1 + 0 & 0 \end{bmatrix} + \begin{bmatrix} -t_i + 0 & x_i + 0 & 0 & 0 \\ y_i + 0 & z_i + 0 & 0 & 0 \\ 0 & 0 & -\lambda + 0 & 0 \\ 0 & 0 & -1 + 0 & 0 \end{bmatrix}.$$

As a block matrix, each $q_i{}^‘$ is composed of the four complex 2 x 2 matrices p_{11}, p_{12}, p_{21}, p_{22} with p_{12} and p_{21} being set to zero.

Now, in order to allow quantum entanglement, the creation operator needs another modification, such that $q_i{}^‘$ x अ© =

$$\begin{bmatrix} 0 - it_i & 0 + ix_i & 0 & 0 \\ 0 + iy_i & 0 + iz_i & 0 & 0 \\ 0 & 0 & 0 + ia_i & 0 + ib_i \\ 0 & 0 & 0 + ic_i & 0 + id_i \end{bmatrix} \times \begin{bmatrix} 0 - 2i & 0 & 0 & 0 \\ 0 & 0 - 2i & 0 & 0 \\ 2 & 0 & 0 - i & 0 \\ 2 & 0 & 0 - i & 0 \end{bmatrix} =$$

$$
\begin{bmatrix}
-2t_i + 0 & 2x_i + 0 & 0 & 0 \\
2y_i + 0 & 2z_i + 0 & 0 & 0 \\
0 + 2i(a_i + b_i) & 0 & +a_i + b_i + 0 & 0 \\
0 + 2i(c_i + d_i) & 0 & +c_i + d_i + 0 & 0
\end{bmatrix} =
$$

$$
\begin{bmatrix}
-t_i + 0 & x_i + 0 & 0 & 0 \\
y_i + 0 & z_i + 0 & 0 & 0 \\
0 + i(a_i + b_i) & 0 & a_i + 0 & 0 \\
0 + i(c_i + d_i) & 0 & c_i + 0 & 0
\end{bmatrix}
+
\begin{bmatrix}
-t_i + 0 & x_i + 0 & 0 & 0 \\
y_i + 0 & z_i + 0 & 0 & 0 \\
0 + i(a_i + b_i) & 0 & b_i + 0 & 0 \\
0 + i(c_i + d_i) & 0 & d_i + 0 & 0
\end{bmatrix} =
$$

with $a_i = +\lambda$, $b_i = -\lambda$ (wavelength) and $c_i = +1$, $d_i = -1$ (spin) or vice versa

$$
\begin{bmatrix}
-t_i + 0 & x_i + 0 & 0 & 0 \\
y_i + 0 & z_i + 0 & 0 & 0 \\
0 + i(\lambda - \lambda) & 0 & +\lambda + 0 & 0 \\
0 + i(1 - 1) & 0 & +1 + 0 & 0
\end{bmatrix}
+
\begin{bmatrix}
-t_i + 0 & x_i + 0 & 0 & 0 \\
y_i + 0 & z_i + 0 & 0 & 0 \\
0 + i(\lambda - \lambda) & 0 & -\lambda + 0 & 0 \\
0 + i(1 - 1) & 0 & -1 + 0 & 0
\end{bmatrix}.
$$

Now, if the spin c_i of particle A changes from +1 to -1,

$i(c_i+d_i)$ turns from $i(1-1) = 0i$ to $i(-1-1) = -2i$,

which will be corrected instantaneously

by a change of d_i from -1 to +1,

so that $i(c_i+d_i)$ turns back to $i(-1+1) = 0i$ and

d_i of particle B changes from -1 to +1,

which means that its spin is now +1 (clockwise).

Hereby, the distance between particle A and B is of no importance, as the correction is performed via matrix representation, within the collective body of the related imaginary matrices $q_i{}'$, which is not limited by the speed of light!

17. Microvita and the Water Rotational Field

Obviously, the grids described in Chapter 14 (Microvita and the Grid) are not enough to support the idea of Deva- and Pretayonis influencing our social and collective life. They provide backgrounds for events in the subatomic realm, but they can hardly do the same for our social affairs: The main problem is the enormous difference in magnitude.

If, however, millions and billions of Microvita are equalized and synchronized as described in Chapter 10 (Interconnectedness of Microvita), they might gain a foothold to our macroscopic world.

Nevertheless, there is another possibility: Transitions from quantum to classical states and reverse had been envisioned as early as 1925, when Albert Einstein - inspired by the young bengali physicist Satyendranath Bose (31) - predicted a strange state of matter (32), later called the Bose-Einstein condensate. According to his calculations, a dilute gas of weakly interacting bosons, cooled down to almost zero degree Kelvin, should show quantum effects on a macroscopic scale.

Seventy years passed until his prediction could be verified experimentally: Eric Cornell, Carl Wieman and co-workers were the first to create a Bose-Einstein condensate (BEC) of approximately two thousand rubidium-87 atoms. Four months later, Wolfgang Ketterle and co-workers independently created a condensate of about two hundred thousand sodium-23 atoms, allowing them to observe quantum mechanical interference from a split condensate (33).

Subsequently, thousands of scientific articles on BEC were published, but until 2006, they all dealt with extremely low temperatures. Overcoming this limitation, it was found that the temperature of Bose-Einstein transition is proportional to $n^{2/3}/m$, where n is the density of the particles and m is their mass, so that the transition temperature can be increased significantly by using dense gases of light quasiparticles in solids. And in 2006, BEC among quasiparticle magnons became possible for the first time at room temperature (34).

Parallel to this, an increasing number of scientists also started considering the possibility of quantum coherence being the ultimate cause of conscious experiences. The approach was pioneered by Alfred Lotka, who proposed as early as 1925 (35) that the mind is able to control the brain by modulating quantum jumps, which would otherwise be either ineffective or productive for random forms of existence only. From there onward, philosophers and nonconformist neurobiologists elaborated on these ideas, later called quantum consciousness, quantum mind or neuroquantology. Their achievements have been summarized by Antonella Vannini (36). Most of them see an amplifying role in local domains of ordered water, fluctuating rapidly between coherent and decoherent rotational states.

In the case at issue, however, the decisive factor is that coherence in the water rotational field is nothing but a means to an end - and the end is to bring the 'charge/form factors' from the imaginary to the macroscopic world: Bose-Einstein condensates of the water rotational field would serve the purpose of amplifying, what has originally been transformed by the described Microvita tensors.

18. Microvita and the Syntropic Principle

In her comparison of the various models of quantum consciousness (36), Antonella Vannini clearly favours the position of Luigi Fantappié: In summary, he affirms a symmetry among entropic and syntropic waves, propagating forwards vs. backwards in time, i.e. diverging from the past to the future vs. diverging from the future to the past (which correspond for us, moving forward in time, to converging waves), thereby acting on present constellations like efficient vs. final causes.

Surprisingly, this stance equates to the position we started with, when I asked in Chapter 1, whether there are 'universalia ante res' impinging on our destiny and fortune - and if so, where such universals could reside - which was answered by defining an operator, allowing time flashes, i.e. a transformation from a past or future imaginary constellation into the real space-time.

Extending this idea to our matrix products

$$q_I{}^\prime \times \text{अ}{}^\prime_{c+} \quad \text{and} \quad q_I{}^\prime \times \text{अ}{}^\prime_{c-}$$

we have to arrange for the following modifications:

$$q_I{}^\prime \times \text{अ}{}^{\prime\oplus}_{c} =$$

$$
\begin{bmatrix}
0 - it_i & 0 + ix_i & 0 & 0 \\
0 + iy_i & 0 + iz_i & 0 & 0 \\
0 & 0 & 0 + ia_i & 0 + ib_i \\
0 & 0 & 0 + ic_i & 0 + id_i
\end{bmatrix}
\times
\begin{bmatrix}
0 - 2iA & 0 & 0 & 0 \\
0 & 0 - 2iA & 0 & 0 \\
0 & 0 & 0 - i & 0 \\
0 & 0 & 0 & 0
\end{bmatrix}
$$

with $t_i = t_r + a$, $\{a \in \mathbb{R} \mid a > 0\}$, $A = t_r / (t_r + a)$ for the syntropic, and

$$q_i' \times \text{अ}_c^{,\ominus} =$$

$$
\begin{bmatrix}
0 - it_i & 0 + ix_i & 0 & 0 \\
0 + iy_i & 0 + iz_i & 0 & 0 \\
0 & 0 & 0 + ia_i & 0 + ib_i \\
0 & 0 & 0 + ic_i & 0 + id_i
\end{bmatrix}
\times
\begin{bmatrix}
0 - 2iA & 0 & 0 & 0 \\
0 & 0 - 2iA & 0 & 0 \\
0 & 0 & 0 & 0 \\
0 & 0 & 0 - i & 0
\end{bmatrix}
$$

with $t_i = t_r + a$, $\{a \in \mathbb{R} \mid -t_r < a < 0\}$, $A = t_r / (t_r + a)$ for the entropic case.

On this basis, the following classes of microvita can be distinguished:

If the positive or negative creation operator performs transformations from the imaginary to the real space in time ($a = 0$, $A = 1$), the matrices are labeled as $\text{अ}_{c+}^{,}$ and $\text{अ}_{c-}^{,}$ respectively.

If the positive creation operator performs transformations from a future imaginary constellation into the present real space-time ($a > 0$, $0 < A < 1$), the matrix is labeled as $\text{अ}_c^{,\oplus}$.

If the negative creation operator performs transformations from a past imaginary constellation into the present real space-time ($-t_r < a < 0$, $A > 1$), the matrix is labeled as $\text{अ}_c^{,\ominus}$.

Due to the described convergence and divergence of waves, अ$_c^{\oplus}$ is generally stronger than अ$_c^{\ominus}$.

Further differences pertain to q_i', particularly the 'charge/form factors' ia_i, ib_i, ic_i and id_i:

According to Fantappié, the *"converging waves are governed by a symmetrical law, which concentrates energy, produces differentiation, structures and order"* (36). Therefrom I conclude that the 'charge/form factors' of the Deva- and Pretayonis, described in Chapter 9, can be understood as a set of more or less symmetrical laws. Consequently, Devayonis produce as much symmetry, as their counterparts, the Pretayonis, produce asymmetry. And among the Devayonis, the laws of Siddha would produce more symmetry than Videhaliina, Videhaliina more than Gandharva, etc., whereas, among the Pretayonis, the laws of Maha'kapala obviously produce more asymmetry than A'kashiipreta, A'kashiipreta more than Brahmadaetya, etc.

These relations are highly suggestive for an analogy with mathematical fractals, which are based on equations undergoing iteration. Some of these equations produce exact self-similarity, others quasi self-similarity, or only statistical self-similarity, overlapping with various degrees of asymmetry. Accordingly, the 'charge/form factors' ia_i, ib_i, ic_i and id_i could correspond to such equations, finally producing the experienced multitude of phenomenal patterns - from the height of beatitude to the abyss of abomination.

19. Conclusion

In Aristotelian science, changes were analysed according to the principle of 'four causes'. Out of these, modern science only accepts material and efficient causes as valid conceptions of change. The other two had been rejected since medieval times, culminating in the young Rudolf Virchow's sentences from 1845 saying that

"the latest medicine has defined its approach as mechanical and its aim as the realisation of a physics of the organisms. It has verified that life is merely an expression of the sum of phenomena, with every one of them proceeding in accordance with common [classical] physics and chemistry (i.e. mechanical laws). It denies the existence of an autocratic force of life and of natural healing" (37).

Nevertheless, this book provides grounds for a rehabilitation of formal and final causes; the proposed microvita theory could indeed help to reintroduce them into an advanced science of the future.

First, the proposed model gives reason on how our known reality could be influenced by imaginary constellations. They can have corresponding space-time coordinates, but they can also be spatially or temporally apart. In the latter case, an influence of an imaginary future equals to a causa finalis. And in the former case, an instant influence over long distances corresponds to telepathic communication.

Second, the influence is not governed by the imaginary constellations, but by microvita, which have been described as specific sets of creation and annihilation operators.

They develop the affinities to the different imaginary matrices, they transform them in singular or in plural, with or without synchronicity. Also the creation operators either split the oxymora into their polarities, or transform only one of them, allowing a differentiation into neutral or positive and negative microvita respectively.

As a result of such transformations we get real objects with defined qualities in space and time; in special cases we get even objects with real as well as imaginary features (see Chapters 3 & 15). And as long as new imaginary opposites (= antithetical factors) can be created, the process will iterate, thereby producing more and more complex forms in real space.

Which means that the process implies creativity, residing in the collective body of imaginary matrices. And as they are all interconnected by immediate matrix representations - this collective body, encompassing the three imaginary times, can be understood as one creative being, otherwise unknown to us: The almost unknown personified.

Glossary

Anujiivat: Sanskrit, anu = of minute size, atomic, jiivat = living-like.

Bosons: Particles with integer spin and totally symmetric composite quantum states, which exempt them from the Pauli exclusion principle.

Brahma: Sanskrit, the combine of Absolute Cognition and its Operative Principle.

Complex Lie groups: Groups with a complex and differentiable manifold, and with the property that the group operations are compatible with a smooth structure.

Determinant: A value computed from the elements of a square matrix by certain equivalent rules. The determinant of a 2×2 matrix A, for example, is defined by $\det(A) = a_{11} a_{22} - a_{12} a_{21}$.

Devayonis: Sanskrit, deva = divine being, yoni = origin; after their demise, elevated human beings become Devayonis = benevolent disembodied souls.

Euclidean space: The coordinate space of all n-tuples of real numbers $(x_1, x_2, ..., x_n)$.

Gell-Mann matrices: A set of generalized Pauli matrices in dimension 3.

Invertible matrix: An n-by-n matrix A, which yields together with an n-by-n matrix B the identity matrix $I = AB = BA$.

Jiiva: Sanskrit, living entity.

Microtubules: Straight, hollow cylinders found throughout the cytoplasm of all eukaryotic cells. They are about 25 nanometers in diameter and between 200 nano- and 25 micrometers in length. They form part of the cytoskeleton that gives structure and shape to a cell, and also serve as conveyor belts moving other organelles throughout the cytoplasm. In addition, microtubules are the major components of cilia and flagella, and participate in the formation of spindle fibers during cell division.

Minkowski space: A four-dimensional real vector space equipped with a symmetric bilinear form η with signature $(-,+,+,+)$ or $(+, -, -, -)$.

Parama Purus'a: Sanskrit, the very Soul of the Universe, Cosmic Cognition.

Prakrti: Sanskrit, Nature.

Pretayonis: Sanskrit, preta = evil being, yoni = origin; after their demise, human beings with a defective mentality become Pretayonis = malevolent disembodied souls.

Quasiparticles: Collective excitations that have particlelike properties such as mass, position and momentum.

Quasiparticle magnons: Quanta of magnetic excitations in a magnetically ordered ensemble of magnetic moments.
Sam'kalpa: Sanskrit, firm determination.

Tanmatras: Sanskrit, tat + matras, tat = that only, matras = essences; the essences of that only, the generic essences of matter.

Three principles of Prakrti: Tendencies or modes of operation, known as sattva, rajas and tamas. Sattva encompasses qualities of goodness, light, and harmony. Rajas is associated with concepts of energy, activity, ambition, and passion; so that, depending on how it is used, it can either have a supportive or hindering effect. Tamas is commonly associated with inertia, darkness, insensitivity and crudity.

Tensor: A mathematical array with an arbitrary number of indices. Insofar tensors are generalizations of scalars (that have no indices), vectors (that have exactly one index), and matrices (that have exactly two indices). They are mainly used in manipulations and transformations of sets of equations within and between different coordinate systems.

Vrajagopa'la: Sanskrit, vraja = joyfully moving along the path, gopa'la = shepherd (a name for Lord Krishna).

References

(1)   Roger Penrose: Shadows of the Mind. 1994

(2)   Wolfram MathWorld: Complex Matrix.
http://mathworld.wolfram.com/ComplexMatrix.html

(3)   Wikipedia: Gell-Mann matrices.
http://en.wikipedia.org/wiki/Gell-Mann_matrices

(4)   Griffiths, D.J.: Introduction to elementary particles. 1987

(5)   Emile Grgin: Structural Unification of Quantum
Mechanics and Relativity. Authorhouse, Indiana, 2007

(6)   Hyperphysics: The Proton. http://hyperphysics.phy-
astr.gsu.edu/hbase/particles/proton.html#c1

(7)   Hyperphysics: The Neutron. http://hyperphysics.phy-
astr.gsu.edu/hbase/particles/proton.html#c3

(8)   The Berkeley Center for Cosmological Physics: Standard
Model of Fundamental Particles and Interactions. http://
bccp.lbl.gov/Academy/workshop08/08%20PDFs/
chart_2006_4.jpg

(9)   Wikipedia: Properties of quarks. http://en.wikipedia.org/wiki/
Quark#Properties_of_quarks

(10) Hyperphysics: The Uncertainty Principle. http://
hyperphysics.phy-astr.gsu.edu/hbase/hframe.html

(11) Hyperphysics: Nuclear Size and Density. http://hyperphysics.phy-astr.gsu.edu/hbase/nuclear/nucuni.html#c4

(12) H.-Y. Cheng: The Proton Spin Puzzle: A Status Report. Chin.J.Phys. 38 (2000) 753

(13) Prabhat Rainjan Sarkar: Microvitum in a Nutshell. 1988, http://shop.anandamarga.org/books/sarkar/mv.htm

(14) Prabhat Rainjan Sarkar: The Goal of Human Ideation. A'nanda Vacana'mrtam, Part 12

(15) Prabhat Rainjan Sarkar: Dances and the Path of Vidya'. A'nanda Vacana'mrtam, Part 12

(16) Prabhat Rainjan Sarkar: The Influence of Ma'ya' on the Human Mind. Tattva Kaomudii, Part 2

(17) Wikipedia: Planck length. http://en.wikipedia.org/wiki/Planck_length

(18) Prabhat Rainjan Sarkar: Microvitum – the Mysterious Emanation of Cosmic Factor

(19) Oliver Hirschbiegel: Der Untergang. 2004, http://www.imdb.de/title/tt0363163/

(20) Richard Bradley: Allied to the Mafia. 1994

(21) Prabhat Rainjan Sarkar: Discourse 13. Shabda Cayanika', Part 2, 1985

(22) Gottfried Wilhelm Leibniz: The Monadology (1714).
translated by Robert Latta (first published in 1898)

(23) Françoise Hudry (ed.): Liber Viginti Quattuor
Philosophorum. Cura et studio F. Hudry, Hermes latinus III,1,
Turnholt 1997 (Corpus Christanorum, Continuatio Mediaevalis
CXLIII A)

(24) Daniel J. Cook, Henry Rosemont Jr.: I. Chinese Opinion
Concerning God. Note 23, Discourse on the Natural Theology
of the Chinese, 1716

(25) Teruaki Nakagomi: Quantum monadology: A world model to
interpret quantum mechanics and relativity. Open Systems &
Information Dynamics, Number 3, 355-378

(26) Teruaki Nakagomi: Quantum monadology: a consistent
world model for consciousness and physics. BioSystems,
Number 69, 27–38

(27) Prabhat Rainjan Sarkar: A'nanda Vacana'mrtam, Part 12.
Ananda Marga Publications, 1979

(28) Frank Wilczek: The Lightness of Being. Basic Books, 2008

(29) Shrii Shrii Anandamurti: Vrajagopa'la and Pariprashna.
Nama'mi Krs'n'a Sundaram, Discourse 22

(30) Albert Einstein, B. Podolsky, and N. Rosen: Can Quantum-
Mechanical Description of Physical Reality Be Considered
Complete? Phys. Rev. 47, 777–780, 1935,
http://prola.aps.org/pdf/PR/v47/i10/p777_1

(31) Satyendranath Bose: Planck's Law and the Light Quantum Hypothesis. Zeitschrift für Physik, 1924, http://www.ias.ac.in/jarch/jaa/15/3-7.pdf

(32) Albert Einstein: Zur Quantentheorie des einatomigen idealen Gases. Sitzungsberichte der Preussischen Akademie der Wissenschaften, I, 1925, http://www.lorentz.leidenuniv.nl/history/Einstein_archive/Einstein_1925_publication/

(33) Wolfgang Ketterle: When Atoms Behave as Waves: Bose-Einstein Condensation and the Atom Laser. Nobel Lecture, 2001, http://nobelprize.org/nobel_prizes/physics/laureates/2001/ketterle-lecture.pdf

(34) Sergej O. Demokritov et al.: Bose–Einstein condensation of quasi-equilibrium magnons at room temperature under pumping. Nature 443, 430-433, 2006, http://www.nature.com/nature/journal/v443/n7110/full/nature05117.html

(35) Alfred James Lotka: Elements of Physical Biology. Williams & Wilkins Co, Baltimore, 1925; reprinted as Elements of Mathematical Biology, Dover Publications, New York, 1956

(36) Antonella Vannini: A Syntropic Model of Consciousness. Thesis, Università Degli Studi di Roma 'La Sapienza' Facoltà di Psicologia 1, 2009, http://www.sintropia.it/thesis.pdf

(37) Rudolf Virchow: Über das Bedürfnis und die Richtigkeit einer Medizin vom mechanischen Standpunkt. Arch Path Anat 7, 188, 1907

Previous Publications

Rudolph, H.-J., K.-D. Gerbitz, G. Michal, and O.H. Wieland: Enzymic Determination of Branched-Chain Amino Acids. Clinical Chemistry 27 (3), pp. 431 - 433 (1981)

Rudolph, H.-J.: Complexities in the Time Domaine: Interlinkage of Time Structures and States of Mind. New Renaissance 2 (1), p. 23 (1991)

Rudolph, H.-J. and M. Bühring: Increase of Heart Rate Variation and Well-Being after External Cold Water Application. European Journal of Physical Medicine & Rehabilitation 2 (4), pp. 102 - 106 (1992)

Rudolph, H.-J.: Respiratory sinus arrhythmia after external cold water application. In: Gutenbrunner C., Hildebrandt G. and Moog R. (eds.), Chronobiology & Chronomedicine, Basic Research and Applications, pp. 310 - 314. Peter Lang, Frankfurt am Main (1993)

Rudolph, H.-J., C. Erben and M. Bühring: Crataegus oxyacantha and the heart rate variability: A placebo-controlled double-blind trial in 20 geriatric patients. Wissenschaftswoche 1992, Forschungsprojekte am Klinikum Steglitz, pp. 150-151 (1992)

Rudolph, H.-J.: Moderne und traditionell-indische Medizin: Eine Initiative der Ärztekammer Berlin. Berliner Ärzte 29 (6), p. 39 (1992)

Rudolph, H.-J.: Traditionelle Indische Medizin. Berliner Ärzte 29 (12), p. 9 (1992)

Rudolph, H.-J.: Der Beauftragte für Indische Medizin. Berliner Ärzte 30 (8), p. 26 (1993)

Rudolph, H.-J.: Time Patterns and the State of Mind. 2nd Gauss Symposium, Proceedings of Biomedical Physics, p. 89 (1993)

Rudolph, H.-J., J. Rohwer, W. Hopfenmüller and M. Bühring: Physical and complementary therapies of scleroderma - results of a questionnaire-based investigation on 411 patients. New Trends in Scleroderma Research - Pathophysiology, Clinics and Therapy, p. 27 (1993)

Rudolph, H.-J.: Ayurvedische Diätetik. Erfahrungsheilkunde 42 (10 a), pp. 558 - 559 (1993)

Rudolph, H.-J.: Paracelsus und die moderne Naturheilkunde. Heilkunst, 106 (12), pp. 26 - 37 (1993)

Rudolph, H.-J.: Paracelsus und die Naturheilkunde. In: Jütte R. (Hrsg.), Entwürfe im Lichte der Natur - Paracelsus heute, pp. 41-54. Karl F. Haug Verlag, Heidelberg (1994)

Rudolph, H.-J.: Wie Weißdorn das Herz stützt. Ärztliche Praxis, 46 (79), pp. 22 - 23 (1994)

Rudolph, H.-J.: Kurze Einführung in die Ayurvedische Diätetik. Erfahrungsheilkunde 43 (8), pp. 445 - 449 (1994)

Rudolph, H.-J.: Traditionelle Indische Medizin. In: Hildebrandt H. (Hrsg.), Pschyrembel Wörterbuch Naturheilkunde und alternative Heilverfahren. Walter de Gruyter, Berlin (1996)

Acknowledgement

Without the help of my spiritual guide, Acarya Vimalananda Avadhuta, as well as the patience and support of my wife Rekha, my son Achill, my daughter Madhuri, and my good friends Gisela Muth, John Williamson, Michael Towsey, Richard Gauthier and Thomas Paasch, this project could have never been realized.